First Steps in Random Walks

From Tools to Applications

J. Klafter and I.M. Sokolov

OXFORD
UNIVERSITY PRESS

Great Clarendon Street, Oxford, OX2 6DP,
United Kingdom

Oxford University Press is a department of the University of Oxford.
It furthers the University's objective of excellence in research, scholarship,
and education by publishing worldwide. Oxford is a registered trade mark of
Oxford University Press in the UK and in certain other countries

First published 2011
First published in paperback 2015

Impression: 1

Published in the United States of America by Oxford University Press
198 Madison Avenue, New York, NY 10016, United States of America

British Library Cataloguing in Publication Data
Data available

Library of Congress Cataloging in Publication Data
Data available

ISBN 978–0–19–923486–8 (Hbk.)
ISBN 978–0–19–875409–1 (Pbk.)

Printed and bound in Great Britain by
Clays Ltd, St Ives plc

Preface

The name "random walk" for the problem of describing the displacement of a point in a sequence of independent random steps appeared when Karl Pearson posed in 1905 a question to the readers of *Nature*: "Can any of your readers refer me to a work wherein I should find a solution of the following problem, or failing the knowledge of any existing solution provide me with an original one? I should be extremely grateful for aid in the matter. A man starts from a point O and walks l yards in a straight line; he then turns through any angle whatever and walks another l yards in a second straight line. He repeats this process n times. I require the probability that after n stretches he is at a distance between r and $r + \delta r$ from his starting point O."

The same year Pearson formulated the problem, a similar problem was formulated in a different context by Albert Einstein in his famous article "The motion of elements suspended in static liquids as claimed in the molecular kinetic theory of heat," which formed part of his PhD work. An independent discussion based on a combinatorial approach was published by Marian Smoluchowski a year later. However, the problem of a random walk was posed even earlier by Louis Bachelier in his thesis devoted to the theory of financial speculations in 1900.

The history of the random walk shows how versatile the model is: Pearson's question was motivated by modeling the motion of animals (not men, but mosquitoes!), a problem he worked on in those days. Einstein's work was clearly pertinent to physics, and Bachelier's to economics (although Bachelier got his PhD in physics for it!). Nowadays, the theory of random walks gives an approach that has proved useful in physics and chemistry (diffusion in different media, reactions, mixing in hydrodynamic flows), economics, biology (from the motion of animals to that of subcellular structures in the crowded environment inside cells), and many other disciplines. Corresponding to a clear geometric and probabilistic model, the random-walk approach is extremely powerful, and can serve not only as a model of simple diffusion, but as a model for many complex sub- and superdiffusive transport processes as well. This book discusses the main variants of the random-walk models and offers the most important mathematical tools for theoretical descriptions of random walks.

Some words about the book should be given here.

Level. The level of the discussion is appropriate for a student starting scientific work or for a scientist entering a new field in which the random-walk approaches are used. Both authors have taught courses on the subject for some years for undergraduate and graduate students in chemistry and physics. The mathematical level of the discussion is therefore that which can be followed after completion of the usual mathematical courses for natural scientists or engineers. Any nonstandard material (such as generating functions or fractional derivatives) is discussed in detail and by use of exercises. In the case of higher levels of mathematical difficulties (especially where the results are not used repeatedly, in which case it would be beneficial to learn something new), we have omitted the derivation and simply stated the result, giving a reference.

Boxes. The most important or the most striking results are given in boxes. These may serve as a short summary of the corresponding discussion.

Exercises. Exercises often contain parts of those derivations that we have omitted in the main discussion, or give additional results. We often refer to the results of the exercises in subsequent chapters. In this case the solution to the exercise is explicitly stated in the text. Some of the exercises were considered to be serious scientific problems in their time. Since the way to a solution is not always trivial, we give hints. If you can solve the exercises, you can really master the subject.

References. This is a self-contained textbook, and therefore very few references are given to the original works. We mostly refer to mathematical compendia (e.g., when we need to evaluate an integral or to sum a series) and to monographs or easily accessible review articles where additional information can be found. The references to original works are given only when they contain additional material that is not discussed in detail in the text. Additional material, or discussions of mathematical problems that may arise in connection with the discussions in the corresponding chapters, are also given in *Further reading.*

Finally, we the authors had fun writing the book. We hope you will have fun using it.

Contents

Abbreviations

CLT	Central Limit Theorem
CTRW	continuous-time random walks
GCLT	generalized Central Limit Theorem
GME	generalized master equation
MSD	mean squared displacement
PDF	probability density function

1
Characteristic functions

> "Life is the sum of trifling motions."
>
> Joseph Brodsky

In this chapter we introduce the characteristic function, a tool that plays a central role in the mathematical description of random walks. When Karl Pearson posed in 1905 his question (reproduced in the Preface) to the readers of *Nature* he was unaware of how simply the problem could be formulated in terms of characteristic functions. A trajectory of the Pearson's random walk is shown in Fig. 1.1.

After getting acquainted with the powerful instrument of characteristic functions, you should be able to answer Pearson's question by performing a short calculation.

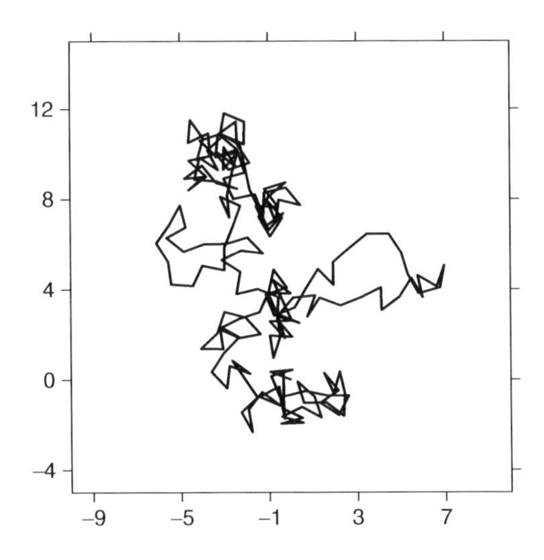

Fig. 1.1 *Pearson's random walk of 200 steps of unit length.*

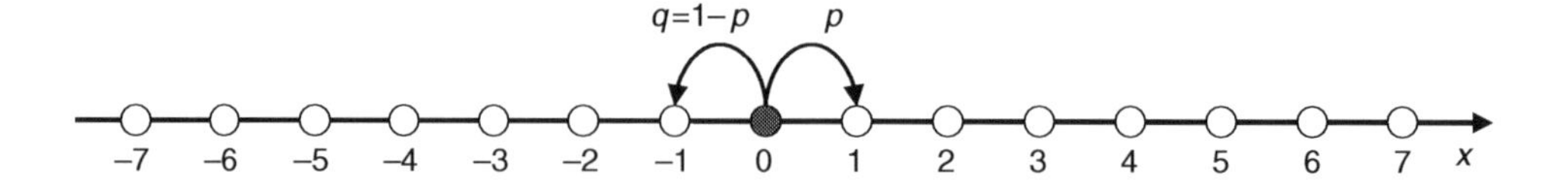

Fig. 1.2 *A schematic illustration of a lattice random-walk problem on a line.*

1.1 First example: A random walk on a one-dimensional lattice

Let us start from a simple example of a random walk on a one-dimensional lattice (Fig. 1.2).

We consider a particle (a "random walker") starting at site 0. Its motion can be described as jumps (steps) to one of the neighboring sites. Different steps of the random walk are considered independent. Steps to the right are performed with probability p and steps to the left with probability $q = 1 - p$. The position of the walker after n steps is determined by the number of steps to the right and to the left. We now introduce an instrument that allows for finding the positions of the walker.

Let us consider the expression $pe^{i\theta} + qe^{-i\theta}$. The coefficient preceding $e^{i\theta}$ is the probability that the first step was to the right, and the coefficient preceding $e^{-i\theta}$ is the probability that the first step was to the left. If we now consider the square of this expression, $(pe^{i\theta} + qe^{-i\theta})^2$, and expand it using the binomial formula $(pe^{i\theta} + qe^{-i\theta})^2 = p^2e^{2i\theta} + 2pq + q^2e^{-2i\theta}$, we see that the coefficient of the first term of the expansion gives the probability that the first two steps were to the right, i.e., that the walker arrives at site $j = 2$ after two steps. The second term corresponds to the probability that the first two steps were made in the opposite direction and that the particle returns to its original position $j = 0$. The coefficient of the third term gives the probability of arriving at $j = -2$. Generalizing these findings to any n, we see that the coefficient preceding $e^{i\theta j}$ in the expansion of the polynomial $(pe^{i\theta} + qe^{-i\theta})^n$ gives us the probability $P_n(j)$ of arriving at site j after n steps. Instead of expanding a polynomial, we can extract $P_n(j)$ from the corresponding expression via Fourier transform:

$$P_n(j) = \frac{1}{2\pi} \int_{-\pi}^{\pi} (pe^{i\theta} + qe^{-i\theta})^n e^{-i\theta j} d\theta. \tag{1.1}$$

The Fourier transform actually "filters out" the coefficient that corresponds to the j-th position. To see this, it is enough to note that $\frac{1}{2\pi} \int_{-\pi}^{\pi} e^{-i\theta n} d\theta = \delta_{n,0}$. The expression $\lambda(\theta) = pe^{i\theta} + qe^{-i\theta}$ is the characteristic function of the distribution of the walker's displacement per step in the discrete one-dimensional random walk. In our example the displacement x takes two values, $x = 1$ and $x = -1$, with probabilities p and q respectively. The characteristic function is therefore the mean $\lambda(\theta) = \langle e^{i\theta x} \rangle$.

As the simplest example, we consider the case of a symmetric random walk, $p = q = \frac{1}{2}$, which has the characteristic function

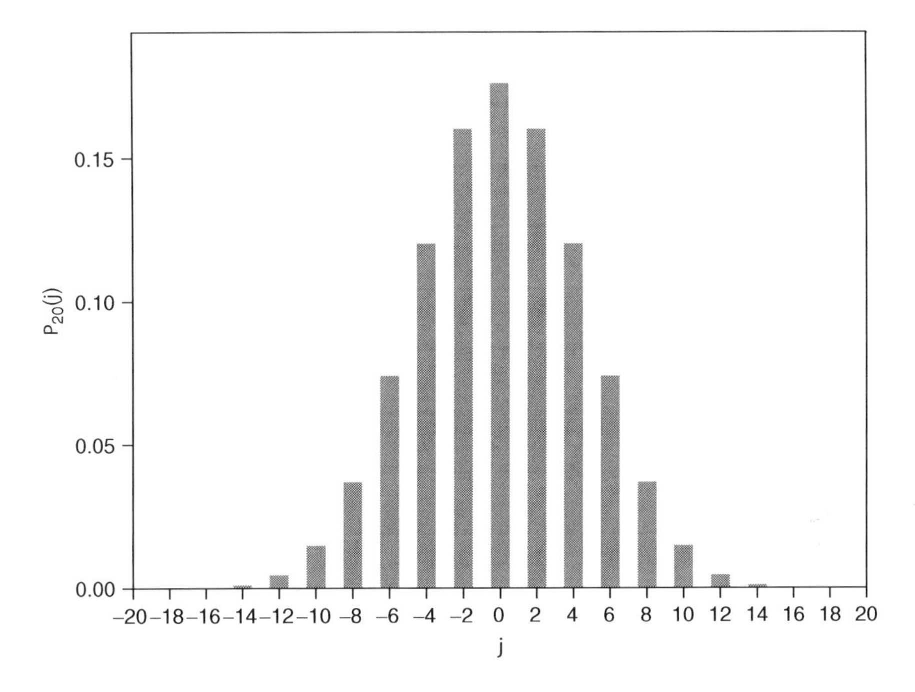

Fig. 1.3 *A histogram of the probability of being at site j after $n = 20$ steps, as given by Eq.(1.2).*

$$\lambda(\theta) = \left(e^{i\theta} + e^{-i\theta}\right)/2 = \cos\theta.$$

The probability of arriving at site j after n steps now reads:

$$\begin{aligned} P_n(j) &= \frac{1}{2\pi} \int_{-\pi}^{\pi} (\cos\theta)^n e^{-i\theta j} d\theta = \frac{1}{2^{n+1}} \left[1 + (-1)^{n+j}\right] \binom{n}{\frac{n+j}{2}}, \\ &= \frac{1}{2^{n+1}} \left[1 + (-1)^{n+j}\right] \frac{n!}{\left(\frac{n+j}{2}\right)! \left(\frac{n-j}{2}\right)!}, \end{aligned} \tag{1.2}$$

which is a result that can be obtained by taking $e^{i\theta j} = \cos(\theta j) + i\sin(\theta j)$ and noting that the integral containing the sine term vanishes, while the one with the cosine term is nonzero and is given by Eq.(3.631.17) of Ref.[1]. Equation (1.2) has two important properties: First, $P_n(j)$ vanishes if n and j have different parities; namely, after an even (odd) number of steps the walker can reside only on a site with an even (odd) number. Second, the probability of obtaining $j > n$ vanishes because of the fact that a factorial of an integer negative number diverges. The displacement of the walker after n steps never exceeds n. Equation (1.2) is the solution to Pearson's problem on a line (see Fig. 1.3).

1.2 More general considerations

Assume that the random walker chooses its displacement per step from a probability density function (PDF) $p(x)$. In this case the characteristic function of the displacement per step is given by

$$\lambda(k) = \langle e^{ikx} \rangle = \int_{-\infty}^{\infty} e^{ikx} p(x) dx, \tag{1.3}$$

where we have taken the argument of the characteristic function to be k instead of θ, which is a standard notation in the continuous case. Mathematically this characteristic function is the Fourier transform of the probability density $p(x)$. Note that one important property of the characteristic function is

$$\lambda(0) = \int_{-\infty}^{\infty} p(x) dx = 1,$$

which follows immediately from its definition and normalization of $p(x)$.

Let us now consider a more general random walk on the line in which the distribution of displacements per step can be continuous, so that the particle's positions are not confined to sites on a lattice. Initially, before performing any step, the particle is situated at $x = 0$; this fact will be mathematically mirrored by taking $P_0(x) = \delta(x)$, with $\delta(x)$ being Dirac's delta-function (δ-function). The probability density of the particle's displacement after the first step is given exactly by $p(x)$, i.e., $P_1(x) = p(x)$. For a random-walk process with independent steps the following recursion relation holds:

$$P_n(x) = \int_{-\infty}^{\infty} P_{n-1}(y) p(x - y) dy. \tag{1.4}$$

The meaning of this relation is as follows: The probability of having a particle at x after n steps is the probability of arriving at y in $n-1$ steps and making up the difference in displacements $x - y$ in one additional step. Mathematically, Eq.(1.4) has the structure of a convolution. Such structures can be easily represented using the Fourier transform. Let us define $P_n(k) = \int_{-\infty}^{\infty} P_n(x) e^{ikx} dx$. Under the Fourier transform the convolution structure of Eq.(1.4) is changed for a simple product

$$P_n(k) = P_{n-1}(k)\lambda(k)$$

so that

$$P_n(k) = P_{n-1}(k)\lambda(k) = P_{n-2}(k)\lambda^2(k) = \ldots = \lambda^n(k), \tag{1.5}$$

which is an exact analog of our $\cos^n(\phi)$ for the discrete case. According to the definition of $P_n(k)$, the probability density $P_n(x)$ can be obtained as an inverse Fourier transform

$$P_n(x) = \frac{1}{2\pi}\int\limits_{-\infty}^{\infty} P_n(k)e^{-ikx}dk = \frac{1}{2\pi}\int\limits_{-\infty}^{\infty} \lambda^n(k)e^{-ikx}dk. \tag{1.6}$$

Equation (1.6) demonstrates that working in the Fourier space leads to simpler structures than in real space: The characteristic function of the distribution of the walker's positions after n steps is simply a product of n characteristic functions of single steps. The simplification introduced by Fourier transform will be further emphasized when turning to more complex problems in the following chapters.

The characteristic function of a sum of n independent, identically distributed random variables is given by the n-th power of the characteristic function of the distribution of each of them:

$$P_n(k) = \lambda^n(k).$$

The corresponding PDF is given by the inverse Fourier transform

$$P_n(x) = \frac{1}{2\pi}\int_{-\infty}^{\infty} \lambda^n(k)e^{-ikx}dk$$

Example 1.1 Let us take the step length distribution as following a Gaussian $p(x) = \frac{1}{\sqrt{2\pi\sigma^2}}e^{-\frac{x^2}{2\sigma^2}}$. Then the characteristic function $\lambda(k)$ reads $\lambda(k) = e^{-\frac{\sigma^2k^2}{2}}$, so that $P_n(k) = \lambda^n(k) = e^{-n\frac{\sigma^2k^2}{2}}$, whose inverse Fourier transform is a Gaussian again: $P_n(x) = \frac{1}{\sqrt{2\pi n\sigma^2}}e^{-\frac{x^2}{2n\sigma^2}}$.

Example 1.2 Let the probability density of displacements in one step be $p(x) = \frac{a}{2}e^{-a|x|}$. In this case $\lambda(k)$ reads $\lambda(k) = \int\limits_{-\infty}^{\infty} p(x)e^{ikx}dx = \frac{a^2}{a^2+k^2}$ and therefore $P_n(x) = \frac{ae^{-a|x|}}{2^{2n-1}(n-1)!}\sum\limits_{m=0}^{n-1}\frac{(2n-2m-2)!(2a|x|)^m}{m!(n-m-1)!}$ (see Ref.[1], Eq.(3.737)).

Exercise 1.1 Calculate the probability density $P_n(x)$ of the displacements after n steps in a random walk, where the step's length is distributed according to a law $p(x) = \frac{b}{\pi}\frac{1}{b^2+x^2}$ (a Cauchy distribution).

Exercise 1.2 Using the expression $P_n(k) = \cos^n(k)$ for the discrete symmetric random walks, show that the corresponding PDF reads $P_n(x) = \sum\limits_j P_n(j)\delta(x-j)$, with $P_n(j)$ given by Eq.(1.2), i.e., that the possible locations of the particle are the points with integer values of the coordinate x, and that the probabilities of finding them there are given exactly by the expressions discussed above.

1.3 The moments

The characteristic function of a distribution is a generating function of its moments. The m-th moment ($m = 1, 2, 3 \ldots$) of a probability distribution of x with density $p(x)$ is defined as

$$M_m = \langle x^m \rangle = \int_{-\infty}^{\infty} x^m p(x) dx.$$

Expanding the exponential in our definition of the characteristic function into a Taylor series we get:

$$\lambda(k) = \int_{-\infty}^{\infty} p(x) \left[1 + ikx - \frac{k^2 x^2}{2} - \frac{ik^3 x^3}{6} + \ldots \right] dx = \sum_{n=0}^{\infty} i^n \frac{M_n}{n!} k^n. \tag{1.7}$$

Therefore

$$M_m = \langle x^m \rangle = (-i)^m \left. \frac{d^m \lambda(k)}{dk^m} \right|_{k=0}. \tag{1.8}$$

The probability distribution does not have to possess all moments. The Cauchy distribution of Exercise 1.1, for example, possesses none. We have to note that, because $|e^{ikx}| = 1$, the integral representing the characteristic function converges for any probability density $p(x)$, while the series of Eq.(1.7) does not have to converge. The existing moments can still be obtained using Eq.(1.8). In many cases connected with the diffusion, we are mainly interested in the second moment of the distribution of the walker's displacement, called the mean squared displacement (MSD).

1.4 Random walk as a process with independent increments

Let us present another point of view on a random walk. Let $x_1, x_2, \ldots, x_n$ be the displacements of the random walker in the 1st, 2nd, ..., n-th step. Then the position of the walker after n steps is given by $X_n = x_1 + x_2 + \ldots + x_n$. It is easy to obtain immediately the characteristic function of the distribution of such positions if we use the definition of the characteristic function as a mean value $\lambda(k) = \left\langle e^{ikx} \right\rangle$ as mentioned earlier. We then have

$$\begin{aligned} P_n(k) &= \left\langle e^{ik(x_1 + x_2 + \ldots + x_n)} \right\rangle = \left\langle e^{ikx_1} e^{ikx_2} \cdot \ldots \cdot e^{ikx_n} \right\rangle \\ &= \left\langle e^{ikx_1} \right\rangle \left\langle e^{ikx_2} \right\rangle \cdot \ldots \cdot \left\langle e^{ikx_n} \right\rangle = \lambda^n(k), \end{aligned}$$

where the possibility of writing the mean value of a product as a product of mean values stems from the independence of the multipliers. This is actually a rederivation of Eq.(1.5).

Let us discuss yet another derivation of this simple relation, again starting from the definition $X_n = x_1 + x_2 + \ldots + x_n$. Reiterating Eq.(1.4) we get

$$P_n(x) = \int_{-\infty}^{\infty} P_{n-1}(y)p(x-y)dy = \ldots = \int_{-\infty}^{\infty} \ldots \int_{-\infty}^{\infty} p(x_1)p(x_2)\ldots p(x-x_{n-1})dx_1 \ldots dx_{n-1},$$

$$= \int_{-\infty}^{\infty} \ldots \int_{-\infty}^{\infty} p(x_1)p(x_2)\ldots p(x_{n-1})p(x_n)\delta\left(\sum_{i=1}^{n} x_i - x\right)dx_1 \ldots dx_n \tag{1.9}$$

where the introduction of the δ-function and of the additional integration serve at first as a formal trick. This gives us another representation of $P_n(x)$, namely, $P_n(x) = \left\langle \delta\left(x - \sum_{i=1}^{n} x_i\right)\right\rangle$. Returning to Eq.(1.9) and using the Fourier representation of the δ-function $\int_{-\infty}^{\infty} \delta(y-x)e^{ikx}dx = e^{iky}$ we get

$$P_n(k) = \int_{-\infty}^{\infty} \ldots \int_{-\infty}^{\infty} p(x_1)p(x_2)\ldots p(x_{n-1})p(x_n)\exp\left(ik\sum_{i=1}^{n} x_i\right)dx_1 \ldots dx_n,$$

$$= \prod_{i=1}^{n}\left[\int_{-\infty}^{\infty} p(x_i)e^{ikx_i}dx_i\right] = [\lambda(k)]^{\,n}$$

where again we use the independence of the multipliers.

1.5 A pedestrian approach to the Central Limit Theorem

As already stated, the probability density of displacement after n steps is given by the inverse Fourier transform of the corresponding characteristic function:

$$P_n(x) = \frac{1}{2\pi}\int_{-\infty}^{\infty} P_n(k)e^{-ikx}dk = \frac{1}{2\pi}\int_{-\infty}^{\infty} [\lambda(k)]^n e^{-ikx}dk.$$

The behavior of $P_n(x)$ at large x is mostly determined by the behavior of $P_n(k)$ at small k. The latter is given by the small-k asymptotics of the $\lambda(k)$. We now consider two examples:

Example 1.3 Let us return to our discrete model in Sec. 1.1 now given in terms of continuous representation with $\lambda(k) = \cos k$. Expanding $\cos k$ up to the second order in k, $\cos k \cong 1 - \frac{k^2}{2} + \ldots$, we get

$$P_n(x) \approx \frac{1}{2\pi}\int_{-\infty}^{\infty}\left[1 - \frac{k^2}{2} + \ldots\right]^n e^{-ikx}dk = \frac{1}{2\pi}\int_{-\infty}^{\infty}\exp\left[n\ln\left(1 - \frac{k^2}{2} + \ldots\right)\right]e^{-ikx}dk \cong,$$

$$\cong \frac{1}{2\pi}\int_{-\infty}^{\infty}\exp\left(-\frac{nk^2}{2}\right)e^{-ikx}dk = \frac{1}{\sqrt{2\pi n}}e^{-\frac{x^2}{2n}} \tag{1.10}$$

namely, a Gaussian distribution.

Example 1.4 Another type of random walk discussed in Example 1.2 is the one with the double-sided exponential density of displacements per step $p(x) = \frac{a}{2}e^{-a|x|}$ and with the characteristic function $\lambda(k) = \frac{a^2}{a^2+k^2} \approx 1 - \frac{k^2}{a^2} + \ldots$. Following the same procedure as in the previous example, we get $P_n(x) \approx \frac{a}{\sqrt{2\pi n}}e^{-\frac{a^2x^2}{2n}}$, namely, a Gaussian distribution again.

From these two examples of very different random walks (one with bounded steps, another with unbounded ones) we learn essentially the following: *Whenever the second moment of the displacement of a step in a symmetric random walk exists so that* $M_2 = \sigma^2$ *(i.e., whenever the characteristic function of a step can be expanded in the form* $\lambda(k) = 1 - \frac{\sigma^2 k^2}{2} + o(k^2)$*), the PDF of the displacements of the walker after n steps tends, for n large enough, to a Gaussian of the form* $P_n(x) = \frac{1}{\sqrt{2\pi n\sigma^2}}\exp\left(-\frac{x^2}{2n\sigma^2}\right)$. This statement provides the content of the Central Limit Theorem (CLT).

Note that if the variance (MSD $\sigma^2 = \langle l^2\rangle$) of a step displacement in a symmetric random walk exists, the MSD after n steps behaves as

$$\langle x^2(n)\rangle = \langle l^2\rangle n. \tag{1.11}$$

The MSD of a random walker after n steps is n times its MSD in one step (provided this exists).

Exercise 1.3 Let us return to an asymmetric random walk with the probability p of moving one step to the right and the probability $q = 1 - p$ of moving to the left. Find the first moment $\mu = M_1$ and the variance $\sigma^2 = M_2 - M_1^2$ of the displacements in one step and the corresponding characteristic function of the displacements. Show that $\lambda(k) = 1 + ik\mu - \frac{M_2 k^2}{2} + \ldots$. Consider the PDF of displacements after $n >> 1$ steps and show that this is given by a Gaussian $P_n(x) = \frac{1}{\sqrt{2\pi n\sigma^2}}\exp\left[-\frac{(x-\mu n)^2}{2n\sigma^2}\right]$.

It might seem that by taking the limit $k \to 0$ in our examples we neglect some information contained in the original distribution of displacement per step; however, it can be demonstrated that this limit appears quite naturally when considering the large number of steps $n \to \infty$. If we shift and rescale our x_i according to $s_i = (x_i - \mu)/\sigma$, the new variables s_i now have zero mean and unit variance. Let us consider now the sum of such variables $S_n = s_1 + s_2 + \ldots + s_n$; its connection to X_n is given by $X_n = (S_n + n\mu)\sigma$. The statement of the CLT is then that the distribution of the *normalized sum* $Z_n = S_n/\sqrt{n}$ converges to a Gaussian with zero mean and unit variance when $n \to \infty$.

Provided the characteristic function of the distribution of the variable x is $\lambda(k) = 1 + ik\mu - \frac{M_2}{2}k^2 + o(k^2) = 1 + ik\mu - \frac{\mu^2+\sigma^2}{2}k^2 + o(k^2)$, then that of $z = s/\sqrt{n}$ reads $\lambda_z(k) = \langle e^{ikz}\rangle = 1 - \frac{k^2}{2n} + o\left(\frac{k^2}{n}\right)$. The characteristic function of the distribution of Z_n for $n \to \infty$ is thus

$$\lim_{n\to\infty}\left[1 - \frac{k^2}{2n} + o\left(\frac{k^2}{n}\right)\right]^n = \exp\left(-\frac{k^2}{2}\right),$$

which is that of a Gaussian distribution with unit dispersion and zero mean. The distribution of X_n can be obtained from that of Z_n by the change in variables. We see that taking large n *automatically* brings us to the small-k limit and that the higher-order corrections (containing moments higher than the second one) do not appear.

In our simplistic formulation of the CLT we stated that the PDF of displacements tends to a Gaussian when n grows, but we did not explain in what sense the one tends to the other, i.e., what we mean when we say that the Gaussian function approximates the real distribution given by $P_n(x)$. To understand this, we return to our discrete example of Sec. 1.1. The Gaussian distribution, being a continuous function of x, does not resemble greatly a "fence" of δ-functions of Exercise 1.2. However, the larger n is, the more δ-functions there are and the closer is the weight $P_n(j)$ of the neighboring ones. The distributions tend to each other in the sense that, if we consider them coarse-grained on some length scale Δx, then the overall probability of finding a walker within a Δx interval given by the exact $P_n(x)$ and by the approximation $P_n^{\text{appr}}(x)$ do coincide, i.e., $\int_x^{x+\Delta x} P_n(x')dx' \to \int_x^{x+\Delta x} P_n^{\text{appr}}(x')dx'$.

Of course, a sensibly chosen bin width Δx (comprising many δ-functions) still has to be much smaller than the overall width of the distribution (which grows with n), and therefore taking large n is necessary. For the definition of the convergence discussed above it is enough that the cumulative distribution functions $F(x) = \int_{-\infty}^{x} p(x')dx'$ of both exact and approximate distributions converge to each other. In the case of continuous $p(x)$ the convergence can also be understood in the sense of convergence of PDFs [2].

The distribution of a sum of n independent, identically distributed random variables possessing the mean μ and the variance σ^2 tends for large n to a Gaussian with the mean μn and the variance $\sigma^2 n$.

Stirling's formula

A very effective approximation for $n!$ is given by

$$n! = \sqrt{2\pi n}\, n^n \exp(-n).$$

This approximation works wonderfully even for $n = 2$ (relative error around 4%), and the relative error decays rapidly when n grows.

Before going on and introducing the generalizations of CLT, we suggest that you consider the following exercise, which shows another way of passing from the discrete situation considered in Sec. 1.1 to a Gaussian distribution.

Exercise 1.4 The limiting Gaussian form of the distribution of displacements in a symmetric random walk can be obtained immediately from Eq.(1.2) by use of Stirling's formula for a factorial (see the Box). Obtain the corresponding expression. Discuss the difference in the prefactor of the expression obtained and of the Gaussian, Eq.(1.10), and its connection to the parity properties of $P_n(j)$.

Hint: Take the logarithm of both parts of Eq.(1.2) and use Stirling's formula to get the approximation for the logarithms of the factorials. Assume $x << n$.

1.6 When the Central Limit Theorem breaks down

Let us consider what happens when the distribution of the displacement in one step lacks the second moment necessary for the CLT to hold. One of the examples is given in Exercise 1.1 where the Cauchy distribution of the displacements in one step was considered: $p(x) = \frac{b}{\pi(b^2+x^2)}$. The second moment of this distribution does not exist since the integral $\int_{-\infty}^{\infty} x^2 p(x)dx$ diverges. Essentially, even the first moment of this distribution exists only in the sense of the principal value. The characteristic function of the Cauchy distribution is $\lambda(k) = \exp(-b|k|)$, and is not differentiable at zero, so that for small k it has the form $\lambda(k) \cong 1 - b|k| + \ldots$, and not $\lambda(k) \cong 1 - \langle x^2 \rangle k^2/2 + \ldots$, which is characteristic of symmetric distributions possessing the second moment. Our result for Exercise 1.1 gives as a distribution of displacements after n steps again a Cauchy distribution $P_n(x) = \frac{nb}{\pi(n^2b^2+x^2)}$, which to no extent tends to a Gaussian when n grows. This shows that the Cauchy distribution shares with the Gaussian one the property that a sum of random variables distributed according to the Cauchy or Gaussian law is distributed according to the Cauchy or Gaussian law again. As we will see later, the Gaussian and the Cauchy distributions are not the only ones to share this property, called *stability*.

Let us consider a situation in which $p(x)$ behaves asymptotically as a power law

$$p(x) \propto \frac{A}{|x|^{1+\alpha}}, \tag{1.12}$$

with $0 < \alpha < 2$ (the Cauchy example corresponded to $\alpha = 1$). The lower bound on α follows from the necessity to guarantee normalization, $\int_{-\infty}^{\infty} p(x)dx = 1$. On the other hand, the cases with $\alpha > 2$ comply with the conditions of the CLT, since the variance exists. We refer throughout to the distributions given by Eq.(1.12) as heavy tailed.

Let us evaluate the characteristic function of the corresponding distribution, confining ourselves to the case of small k. To do this we use the following trick:

$$\lambda(k) = 1 - (1 - \lambda(k)) = 1 - \int_{-\infty}^{\infty} (1 - \cos kx)p(x)dx, \tag{1.13}$$

(the *sin*-part of the Fourier transform vanishes due to the symmetry). Contrary to our examples in Sec. 1.3, we cannot expand the cosine since the second moment diverges for $\alpha < 2$. By changing the integration variable to $y = kx$, we get

$$\int_{-\infty}^{\infty} (1 - \cos kx) p(x) dx = \frac{1}{k} \int_{-\infty}^{\infty} (1 - \cos y) p\left(\frac{y}{k}\right) dy \simeq \frac{1}{k} \int_{-\infty}^{\infty} (1 - \cos y) \frac{A}{|y/k|^{1+\alpha}} dy ,$$

$$= A\,|k|^{\alpha} \int_{-\infty}^{\infty} \frac{(1 - \cos y)}{|y|^{1+\alpha}} dy = \tilde{A}\,|k|^{\alpha} .$$

The integral in the last line converges for $\alpha < 2$ and is equal to

$$\int_{-\infty}^{\infty} \frac{(1 - \cos y)}{|y|^{1+\alpha}} dy = \frac{\pi}{\Gamma(1 + \alpha) \sin(\pi\alpha/2)}$$

(the expression can be obtained by transforming Eq.(3.823) in Ref.[1]).

Therefore

$$\lambda(k) = 1 - \tilde{A}\,|k|^{\alpha} + \ldots . \tag{1.14}$$

We stress that Eq.(1.14) holds only for $0 < \alpha < 2$. Imagine that an expression like Eq.(1.14) with $\alpha > 2$ is a characteristic function of the distribution with the PDF $p(x)$. Since this $\lambda(k)$ has two first derivatives that vanish, the two first moments of $p(x)$ are zero: $\int_{-\infty}^{\infty} x p(x) dx = 0$ and $\int_{-\infty}^{\infty} x^2 p(x) dx = 0$. The only distribution possessing the second property is a δ-function, whose characteristic function is, however, not given by Eq.(1.14). Thus for the PDFs $p(x)$ given by Eq.(1.12) with $\alpha > 2$, the expansion of $\lambda(k)$ always contains the quadratic term corresponding to the existence of the second moment $\lambda(k) = 1 - a_2 k^2 + o(k^2)$.

Using the same approach as in previous sections, we can obtain the characteristic function of the walker's positions after n steps, which for large n tends to

$$P_n(k) = \left(1 - \tilde{A}\,|k|^{\alpha} + \ldots\right)^n \to \exp\left(-n\tilde{A}\,|k|^{\alpha}\right) .$$

The characteristic functions of the form $\lambda(k) = \exp(-a|k|^{\alpha})$ with $0 < \alpha < 2$ are the characteristic functions of so-called symmetric Lévy distributions, the Cauchy distribution of Exercise 1.1 being one of them. The Gaussian distribution with the characteristic function $\lambda(k) = \exp(-ak^2)$ can be considered as a limiting case, namely, the only one possessing the second moment. The Lévy distributions will be considered in more detail in Chapter 7 of this book. The parameters a and α in the characteristic function of these distributions are called the scale parameter and the index of the distributions, respectively. All such distributions share the property of stability.

Although the requirements of the CLT break down, we can formulate the analogous theorem pertinent to probability densities displaying asymptotically the power-law behavior as given by Eq.(1.12). As in the case of CLT it is convenient to rescale the random variables x_i by introducing $z = x_i/n^{1/\alpha}$ and to consider the normalized sum $Z_n = \sum_{i=1}^{n} z_i$. The characteristic function of the distribution of $z = x/n^{1/\alpha}$ then is $\lambda_z(k) = 1 - \frac{|k|^\alpha}{n} + o\left(\frac{|k|^\alpha}{n}\right)$. The characteristic function of the distribution of Z_n for $n \to \infty$ is thus

$$\lim_{n\to\infty}\left[1 - \frac{|k|^\alpha}{n} + o\left(\frac{|k|^\alpha}{n}\right)\right]^n = \exp\left(-|k|^\alpha\right),$$

i.e., tends to a characteristic function of the normalized symmetric Lévy distribution. Therefore the distributions of the sums of many independent, symmetric, and identically distributed random variables whose probability densities show power-law tails described by Eq.(1.12) tend to the corresponding symmetric Lévy law in the same sense as the distributions of the variables possessing the second moments tend to a Gaussian. This statement is one of the generalizations of the CLT. We shall refer to this statement as a Generalized Central Limit Theorem (GCLT). Since (like CLT) the GCLT relies on the properties of the characteristic function of the distribution of displacements in one step, it can also be used for lattice walks, with the same interpretation of convergence as in the case of CLT.

Let us now consider an example interpolating between the two types of behavior, i.e., between that of the CLT and the GCLT. We consider a lattice walk, except now we allow the walker to jump not only to the nearest neighbors but also to distant sites, with the probability depending on the distance. We assume

$$p(l) = \frac{a-1}{2a}\sum_{j=0}^{\infty} a^{-j}\left(\delta_{l,b^j} + \delta_{-l,b^j}\right), \tag{1.15}$$

where $a > 1$ is a real number and b is a natural number ($b = 1, 2, 3, \ldots$). The expression in Eq.(1.15) represents a so-called Weierstrass function. The case $b = 1$ corresponds to a simple lattice walk in our first example of Sec. 1.1. Formally, the second moment of the distribution is given by

$$\sum_{l=-\infty}^{\infty} l^2 p(l) = \frac{a-1}{a}\sum_{j=0}^{\infty}\left(b^2/a\right)^j.$$

Depending on the values of a and b the corresponding series can either converge or diverge. For $b^2/a < 1$ the series converges to $\langle l^2\rangle = \frac{a-1}{a-b^2}$, while for $b^2/a \geq 1$ it diverges and the second moment is absent. We proceed to show that in this second case $p(l)$ belongs to the domain of the applicability of GCLT.

The characteristic function of the distribution (1.15) is given by

$$\lambda(k) = \sum_l p(l)e^{ikl} = \frac{a-1}{a}\sum_{j=0}^{\infty} a^{-j}\cos(b^j k). \tag{1.16}$$

For the first case, $b^2/a < 1$, its behavior for $k \to 0$ is given by $\lambda(k) \cong 1 - k^2\langle l^2\rangle/2 + \ldots$. In the second case, $b^2/a \geq 1$, we have to circumvent the explicit expansion of Eq.(1.16) by first showing that the characteristic function Eq.(1.16) fulfills the condition

$$\lambda(bk) = a\lambda(k) - (a-1)\cos k, \tag{1.17}$$

which leads to $\lambda(k) \cong 1 - |k|^\alpha$ with $\alpha = \ln a/\ln b$.

Exercise 1.5 Prove the validity of Eq.(1.17). Show that for $k \to 0$ Eq.(1.17) implies that $1 - p(k) \propto |k|^\alpha$ with $\alpha = \ln a/\ln b$.

Thus, the distribution of the displacements of a Weierstrass random walker tends either to a Gaussian (for $b^2/a < 1$) or to a Lévy distribution with index $\alpha = \ln a/\ln b$ for $b^2/a \geq 1$ and bridges the two types of behavior, which seemed to be quite distinct.

1.7 Random walks in higher dimensions

The results of the previous sections can be generalized to higher dimensions. We again start from the displacement of a walker in one step, which is given by $p(\mathbf{r}) = p(x_1, \ldots, x_d)$ where $(x_1, \ldots, x_d)$ denote the Cartesian coordinates of the corresponding point in a d-dimensional space. The corresponding characteristic function is defined as

$$\lambda(\mathbf{k}) = \left\langle e^{i\mathbf{kr}} \right\rangle = \int_{-\infty}^{\infty} \cdots \int_{-\infty}^{\infty} p(\mathbf{r}) e^{i\mathbf{kr}} d^d\mathbf{r}. \tag{1.18}$$

For the lattice walk we have $p(\mathbf{j}) = p(j_1, \ldots, j_D)$ with the corresponding characteristic function

$$\lambda(\mathbf{\Theta}) = \left\langle e^{i\mathbf{\Theta r}} \right\rangle = \sum_{j_1} \sum_{j_2} \cdots \sum_{j_d} p(\mathbf{j}) e^{i\mathbf{\Theta j}}.$$

The functions $\lambda(\mathbf{\Theta})$ (or $\lambda(\mathbf{k})$ in the continuous case) are the analogs of the characteristic function of the displacements in one step $\lambda(\theta)$ or $\lambda(k)$ in our previous considerations. As in the one-dimensional case, the characteristic function of the displacement's distribution after n steps is $[\lambda(\mathbf{\Theta})]^n$, and the probability of being at site n is given by

$$P_n(\mathbf{j}) = \frac{1}{(2\pi)^d} \int_{-\pi}^{\pi} \cdots \int_{-\pi}^{\pi} \lambda^n(\mathbf{\Theta}) e^{-i\mathbf{\Theta j}} d^d\mathbf{\Theta}$$

in the discrete case and by

$$P_n(\mathbf{x}) = \frac{1}{(2\pi)^d} \int_{-\infty}^{\infty} \cdots \int_{-\infty}^{\infty} \lambda^n(\mathbf{k}) e^{-i\mathbf{kx}} d^d\mathbf{k}$$

in the continuous case.

Exercise 1.6 Evaluate the characteristic function $\lambda(\mathbf{\Theta})$ for the random walk on a square lattice with unit lattice spacing. Show that $\lambda(\mathbf{\Theta}) = \frac{1}{2}[\cos\Theta_x + \cos\Theta_y]$.

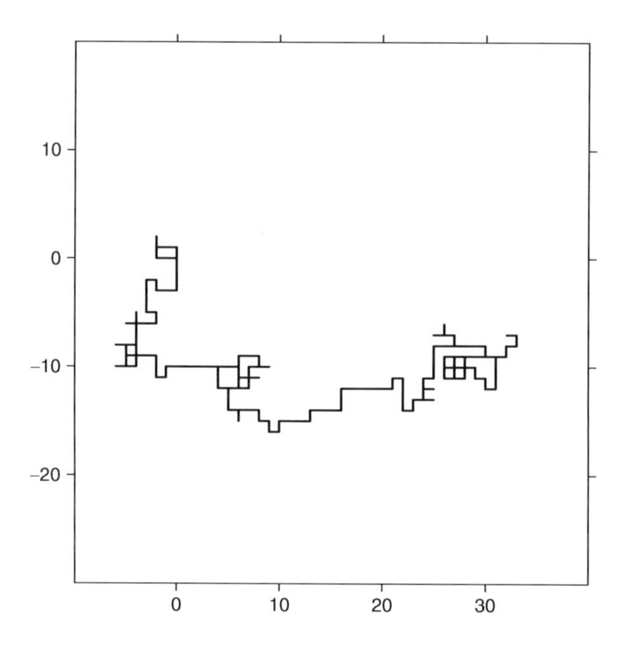

Fig. 1.4 *A lattice random walk with* $n = 200$ *steps in two dimensions.*

A trajectory of such a random walk is given in Fig. 1.4.

Exercise 1.7 Pearson's problem: Consider a two-dimensional random walk with $p(\mathbf{r}) = \frac{1}{2\pi l}\delta(r-l)$, i.e., a walk in which the length of the step is fixed and the direction of the step is randomly chosen. Evaluate the characteristic function $\lambda(\mathbf{k})$ and the asymptotic form of the probability density $P_n(\mathbf{r})$.

Exercise 1.8 Consider a two-dimensional random walk with $p(\mathbf{r}) = \frac{1}{2\pi r}\exp(-r)$, i.e., a walk in which the length of the step is exponentially distributed and the direction of the step is randomly chosen. Evaluate the characteristic function $\lambda(\mathbf{k})$ and the asymptotic form of the probability density $P_n(\mathbf{r})$.

Exercise 1.9 Consider a two-dimensional random walk with $p(\mathbf{r}) = \frac{1}{\pi^2 r(1+r^2)}$, i.e., a walk in which the length of steps is distributed according to the Cauchy law and the direction of the step is randomly chosen. Evaluate the characteristic function $\lambda(\mathbf{k})$ and the asymptotic form of $P_n(\mathbf{r})$.

Hints: Change to polar coordinates. Note that $\frac{1}{2\pi}\int_0^{2\pi} e^{ix\cos\theta}d\theta = J_0(x)$ with $J_0(x)$ being a Bessel function. For the inverse transform use the fact that $\int_0^{\infty} e^{-\alpha x}J_0(\beta x)dx = \frac{1}{\sqrt{\alpha^2+\beta^2}}$ (see Eq.(6.611.1) in Ref.[1]).

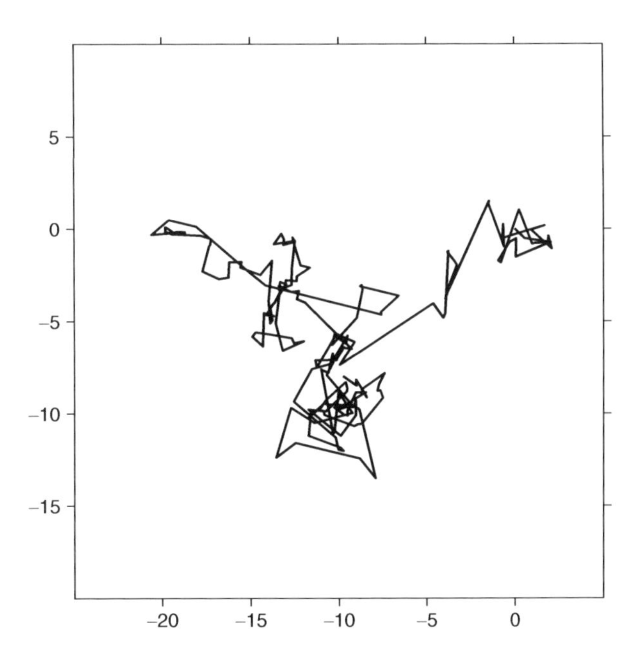

Fig. 1.5 *A random walk with exponential distribution of step lengths and with random angles;* $n = 200$.

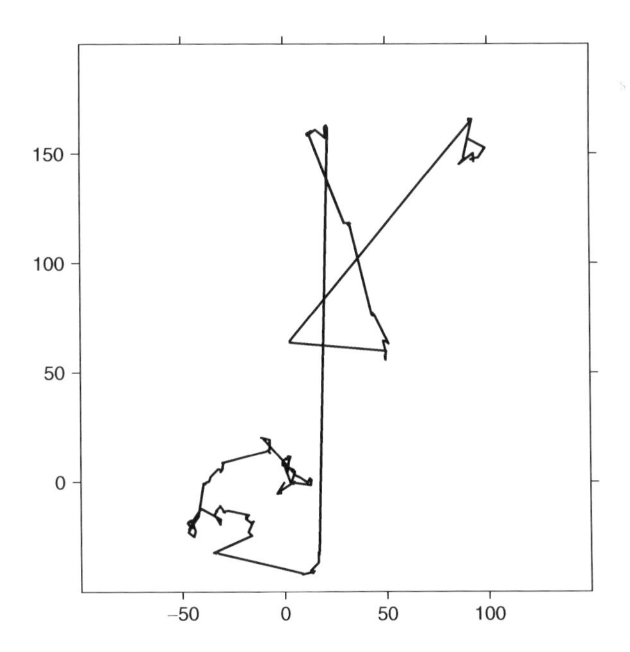

Fig. 1.6 *A random walk with Cauchy distribution of step lengths and with random angles;* $n = 200$. *Note a large difference in scales with Fig. 1.5.*

The sample trajectories of the random walks of Exercises 1.8 and 1.9 are presented in Figs. 1.5 and 1.6.

References

[1] I.S. Gradstein and I.M. Ryzhik. *Table of Integrals, Series and Products*, Boston: Academic Press, 1994
[2] W. Feller. *An Introduction to Probability Theory and Its Applications*, New York: Wiley, 1971 (Random walks are discussed in Ch. III of Vol. 1 in a very different manner to here; the characteristic functions are first introduced in Ch. XV in Vol. 2.)

Further reading

G.H. Weiss. *Aspects and Applications of the Random Walk*, Amsterdam: North-Holland, 1994
B.D. Hughes. *Random Walks and Random Environments*, Vol. 1: *Random Walks*, Oxford: Clarendon, 1996

2
Generating functions and applications

"Let these describe the indescribable."

Lord Byron

In the previous chapter we introduced the concept of characteristic functions and the corresponding moments, whenever they exist, of probability distributions that underlie random walks. However, in many situations we are interested in further information that is not directly derivable from the moments, such as the first-passage and return probabilities of a random walker. As we know already, the random-walk process is characterized by a sequence of probabilities $P_n(\mathbf{r})$ of a random walker's displacements after n steps. A general way to obtain information from such sequences is in terms of the so-called generating functions.

2.1 Definitions and properties

A sequence of numbers $\{f_n\}$ can be considered as the coefficients of a Taylor expansion of a function $f(z)$, which is then called the *generating function* of that sequence,

$$f(z) = \sum_n f_n z^n. \tag{2.1}$$

We assume that the corresponding Taylor series converges in some interval $-z_0 < z < z_0$ on the real axis. If we fortunately know that $\{f_n\}$ are the Taylor coefficients of some tabulated function, then we have a closed form of the function $f(z)$. Some simple examples are:

- An infinite geometric series $1/(1-z)$ corresponding to $f_n = 1$ for all $n \geq 0$,

$$1/(1-z) = 1 + z + z^2 + \ldots.$$

- An exponential $\exp(az)$ corresponding to $f_n = a^n/n!$

$$\exp(az) = \sum_{N=0}^{\infty} \frac{a^n}{n!} z^n.$$

- A binomial $(q - pz)^N$ corresponding to $f_n = \binom{N}{n} p^n q^{N-n}$, $0 \le n \le N$.

Generating functions of number sequences have the following important properties:

- The generating function of the sum of two sequences $h_n = f_n + g_n$ is the sum of the generating functions of each of them: $h(z) = f(z) + g(z)$ (a *sum rule*).
- The generating function of the *convolution* of the two sequences, i.e., of the sequence with the elements $h_n = \sum_{k=0}^{n} f_k g_{n-k}$, is the product of the corresponding generating functions: $h(z) = f(z)g(z)$ (a *convolution rule*).

Exercise 2.1 Check the convolution rule by comparing the coefficients preceding the corresponding powers of z in the expansion of $h(z)$.

Generating functions allow for compactly representing the whole sequence and are an alternative to explicitly writing down the terms. Moreover, the properties above allow us to apply simple algebra to the generating functions instead of manipulating the sequences themselves.

The generating function of a sequence is closely related to the Laplace transform. Imagine that $f_n = \phi(n)$ possess some properties of regularity as a function of n and that the series for $f(z)$ converges in some range of z. In this case the transformation leading from the sequence $\{f_n\}$ to its generating function $f(z)$ (which we here call *z-transform*) is a close discrete analog of the Laplace transform: Take $z = \exp(-u)$ and we recognize in Eq.(2.1) an integral sum for the corresponding Laplace integral $\mathcal{L}\{\phi(x)\} = \int_0^\infty \phi(x)e^{-ux}\,dx$ (insert x instead of n and consider the continuous limit). Thus, under regularity conditions the generating function can be approximated by a Laplace transform. The advantage here is that the tables of Laplace transforms are readily available and applicable; see, e.g., [1–3]. We shall return to the Laplace transform in the next chapters.

Example 2.1 Let us consider the following example from the probability theory. A probability distribution of the results of throwing a die with equally probable outcomes of 1, 2, 3, 4, 5, and 6 (with probability $f_n = P_n = 1/6$ each) corresponds to a generating function

$$f(z) = \frac{1}{6}z + \frac{1}{6}z^2 + \frac{1}{6}z^3 + \frac{1}{6}z^4 + \frac{1}{6}z^5 + \frac{1}{6}z^6 = \frac{1}{6}\frac{z - z^7}{1 - z}. \tag{2.2}$$

Here the coefficients f_n are the probabilities of mutually excluding outcomes. Then

$$f(z)|_{z=1} = \sum_n f_n = \sum_n P_n = 1.$$

Exercise 2.2 Check this normalization condition for Eq.(2.2) using l'Hôpital's rule.

The moments of the corresponding distribution, defined by

$$\langle n^k \rangle = \sum_n n^k P_n,$$

are given by the derivatives of the generating function. For example, the mean value $\langle n \rangle$ is given by

$$\langle n \rangle = \sum_n n f_n = \sum_n n z^{n-1} f_n \bigg|_{z=1} = \frac{d}{dz} f(z) \bigg|_{z=1}.$$

In our example of throwing a die, this mean value is clearly $7/2 = 3.5$. In what follows we will denote such derivatives as $\frac{d}{dz} f(z)\big|_{z=1} = f'(1)$, and in general $\frac{d^n}{dz^n} f(z)\big|_{z=1} = f^{(n)}(1)$. The second moment is then

$$\begin{aligned}\langle n^2 \rangle &= \sum_n n^2 f_n = \sum_n n(n-1) z^{n-2} f_n \bigg|_{z=1} + \sum_n n z^{n-1} f_n \bigg|_{z=1} \\ &= \frac{d^2}{dz^2} f(z) \bigg|_{z=1} + \frac{d}{dz} f(z)\Big|_{z=1} = f''(1) + f'(1).\end{aligned}$$

The variance of the results $\sigma^2 = \langle n^2 \rangle - \langle n \rangle^2$ is equal to

$$\sigma^2 = f''(1) + f'(1) - [f'(1)]^2.$$

If $f(z)$ is the generating function of a probability distribution P_n then:

$$\begin{aligned} f(1) &= 1 \\ \langle n \rangle &= f'(1) \\ \langle n^2 \rangle &= f''(1) + f'(1). \end{aligned}$$

The moments of the distribution are easily obtained from the derivatives of the generating function.

2.2 Tauberian theorems

We can find generating functions of sequences or sequences from their generating functions by using tables or by approximating the z-transform by a Laplace transform. However, for a special class of functions, those which asymptotically behave as power laws, there is no need to do so. Their asymptotic behavior is given by a very simple rule following from Tauberian theorems [4].

Consider a generating function of a sequence $f(z) = \sum_n f_n z^n$ with $f_n > 0$, which behaves essentially as a power law, so that for z close to unity ($z \to 1$)

$$f(z) \cong \frac{1}{(1-z)^\gamma} L\left(\frac{1}{1-z}\right). \tag{2.4}$$

Here γ is some positive number and $L(x)$ is a *slowly varying function* of x, i.e.,

$$\lim_{x\to\infty} \frac{L(Cx)}{L(x)} = 1$$

for any positive constant C. Examples of slowly varying functions are not only functions tending to a finite limit when $x \to \infty$, but also $\log x$ or any power of $\log x$. The Tauberian theorem for sequences states that the partial sum of a series whose generating function fulfills Eq.(2.4) behaves as

$$f_1 + f_2 + \ldots + f_n \cong \frac{1}{\Gamma(\gamma+1)} n^{\gamma} L(n),$$

where $\Gamma(x)$ is the Γ-function [5], and $L(x)$ is the same slowly varying function as in Eq.(2.4). If the sequence $\{f_n\}$ is monotonic (at least starting from some value of n) then

$$f_n \cong \frac{1}{\Gamma(\gamma)} n^{\gamma-1} L(n). \tag{2.5}$$

The quality of the approximation is typically very good. The inverse theorem also holds [4].

Exercise 2.3

a) Using the Tauberian theorem, find the asymptotic form of the sequence whose generating function is

$$g(z) = \frac{1}{1-z}.$$

b) Consider a slightly more complicated case

$$g(z) = \frac{1}{1-z} \ln^3 \frac{1}{1-z}.$$

For monotonic sequences

$$f(z) \cong \frac{1}{(1-z)^{\gamma}} L\left(\frac{1}{1-z}\right)$$
$$\Updownarrow$$
$$f_n \cong \frac{1}{\Gamma(\gamma)} n^{\gamma-1} L(n)$$

Example 2.2 In Sect. 2.3, where we calculate a property called return probability, we encounter the generating function

$$f(z) = 1 - \sqrt{1 - z^2},$$

which appears quite often when describing one-dimensional random walks. This function is simple enough (of a binomial type) to obtain a closed expression for the corresponding sequence. However, it is also a good example for using the above Tauberian theorem. The problem here is that the function itself does not have the form of Eq.(2.4). On the other hand, its derivative

$$g(z) = f'(z) = \frac{z}{\sqrt{1-z^2}}$$

is definitely of the form of Eq.(2.4). For $z \to 1$ it behaves as

$$g(z) \cong \frac{1}{\sqrt{2(1-z)}},$$

which corresponds to $\gamma = 1/2$ and $L(x) = 1/\sqrt{2}$. We note that $g(z)$ is the generating function of the sequence $\{g_n\} = \{nf_n\}$. Now applying the Tauberian theorem, we get

$$nf_n \cong \frac{1}{\sqrt{2}} \frac{1}{\Gamma(1/2)} n^{-1/2}$$

so that

$$f_n \cong \frac{1}{\sqrt{2\pi}} n^{-3/2}$$

(note that $\Gamma(1/2) = \sqrt{\pi}$).[1]

Exercise 2.4 Find the asymptotic form of the sequence whose generating function is

$$g(z) \cong 1 - \sqrt{1-z}$$

as it appears in the derivation of the Sparre Andersen theorem (see Sec. 2.5).

2.3 Application to random walks: The first-passage and return probabilities

We now apply the generating-functions tool to the problem of random walks on a lattice. In Chapter 1 we discussed the probability $P_n(\mathbf{r})$ for a random walker's displacement $\mathbf{r}$ after n steps. If the walker starts at the origin (site $\mathbf{0}$) the same $P_n(\mathbf{r})$ gives us the probability of being at site $\mathbf{r}$ after n steps.

[1] Here we proceed in a way that would be appropriate if the sequence f_n were monotonic, while in reality it is not, see Eq.(2.14) below: all the elements with odd numbers vanish. Our result here delivers an average of subsequent even and odd elements. To understand the situation the reader might compare the sequences generated by functions $g(z)$ and $g(z^2)$ and use the result of Exercise 2.4.

Note that the corresponding probabilities for n and $n-1$ steps are connected by the equation

$$P_n(\mathbf{r}) = \sum_{\mathbf{r}'} p(\mathbf{r},\mathbf{r}')P_{n-1}(\mathbf{r}'), \tag{2.6}$$

where $p(\mathbf{r},\mathbf{r}')$ is the probability of taking a step from the site $\mathbf{r}'$ to site $\mathbf{r}$. On homogeneous lattices the corresponding transition probabilities depend only on $\mathbf{r}-\mathbf{r}'$: $p(\mathbf{r},\mathbf{r}') = p(\mathbf{r}-\mathbf{r}')$. The system of equations (2.6) has to be augmented by an initial condition $P_0(\mathbf{r}) = \delta_{\mathbf{r},0}$. Multiplying both sides of Eq.(2.6) by z^n and summing up all expressions for different n, we arrive at

$$P(\mathbf{r},z) - z\sum_{\mathbf{r}'} p(\mathbf{r}-\mathbf{r}')P(\mathbf{r}',z) = \delta_{\mathbf{r},0}, \tag{2.7}$$

which plays an important role in the discussion of the generalized master equation in Chapter 5.

Exercise 2.5 Show that for the random walk on a hypercubic lattice in d dimensions Eq.(2.7) leads to

$$P(\mathbf{r},z) = \frac{1}{(2\pi)^d}\int_{-\pi}^{\pi}\cdots\int_{-\pi}^{\pi}\frac{e^{i\mathbf{\Theta}\mathbf{r}}}{1-z\lambda(\mathbf{\Theta})}d^d\mathbf{\Theta}.$$

Let us now discuss the probability $F_n(\mathbf{r})$ of visiting site $\mathbf{r}$ *for the first time* at step n. The probabilities $P_n(\mathbf{r})$ and $F_n(\mathbf{r})$ are connected by the following relation:

$$P_n(\mathbf{r}) = \delta_{n,0}\delta_{\mathbf{r},0} + \sum_{k=1}^{n} F_k(\mathbf{r})P_{n-k}(\mathbf{0}). \tag{2.8}$$

The idea is: The walker starts from the origin at step 0 (this fact is represented by the term with δ-functions on the right-hand side) and ends at step n at its final position $\mathbf{r}$. On its way from $\mathbf{0}$ to $\mathbf{r}$ it may have already visited $\mathbf{r}$ at some earlier step k (with probability given by $F_k(\mathbf{r})$), left it, and returned to it during the remaining $n-k$ steps (probability $P_{n-k}(\mathbf{r})$) (see Fig. 2.1).

Now, we turn from the relation in Eq.(2.8) to the corresponding generating functions. To do this we multiply Eq.(2.6) for different n by z^n and add them all together. We get:

$$\sum_n P_n(\mathbf{r})z^n = \delta_{\mathbf{r},0}z^0 + \sum_n z^n \sum_{k=1}^{n} F_k(\mathbf{r})P_{n-k}(\mathbf{0}).$$

Using the definitions of the z-transforms $P(\mathbf{r},z) = \sum_n P_n(\mathbf{r})z^n$ and $F(\mathbf{r},z) = \sum_n F_n(\mathbf{r})z^n$ and applying the convolution rule, we get:

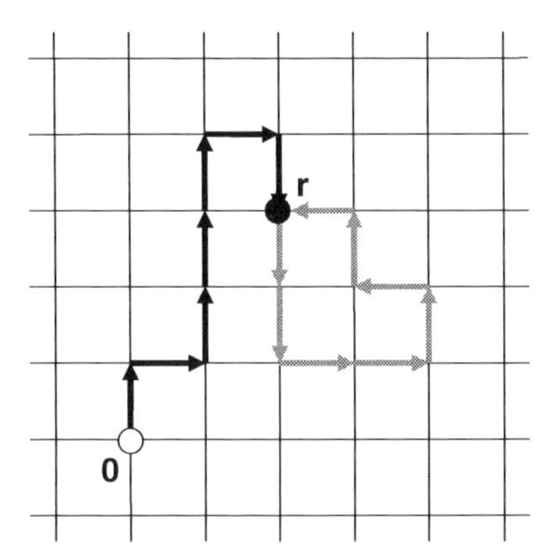

Fig. 2.1 *The trajectory of a random walker starting at **0** and ending at **r**. It consists of a first-passage part from **0** to **r** (black) and of a part leaving **r** and returning (gray).*

$$P(\mathbf{r}, z) = \delta_{\mathbf{r},0} + F(\mathbf{r}, z)P(\mathbf{0}, z),$$

which gives:

$$F(\mathbf{r}, z) = \frac{P(\mathbf{r}, z) - \delta_{\mathbf{r},0}}{P(\mathbf{0}, z)}. \tag{2.9}$$

Equation (2.9) is an important result. Using the information about $P_n(\mathbf{r})$ (Chapter 1), we derived a new property that is connected to $P_N(\mathbf{r})$ in a way that is not obvious.

2.3.1 The return probability

Since $F_n(\mathbf{r})$ is the probability of visiting $\mathbf{r}$ for the first time at step $n \geq 1$, $F_n(\mathbf{0})$ is the probability of the first return to the origin at step n. The overall probability of returning (denoted by $F(\mathbf{0})$) is the sum of the probabilities of first returns at steps $1,\ 2,\ \ldots,\ n,\ \ldots$, i.e.,

$$F(\mathbf{0}) = \sum_{n=0}^{\infty} F_n(\mathbf{0}) \tag{2.10}$$

(note that according to Eq.(2.9) $F_0(\mathbf{0}) = F(\mathbf{0}, z \to 0) = 0$). If $F(\mathbf{0}) < 1$ the random walker does not necessarily return to its starting point (such random walks are called *transient* or *non-recurrent*); on the other hand, if $F(\mathbf{0}) = 1$ the walk is *recurrent.* In an infinitely long recurrent walk the random walker visits any site infinitely many times. The property of recurrence or non-recurrence is of primary importance for chemical reactions and related search theories.

Recalling that $F(\mathbf{r}, z) = \sum_n F_n(\mathbf{r}) z^n$, it is clear that $F(\mathbf{0}) = \sum_n F_n(\mathbf{0}) z^n \Big|_{z=1} = F(\mathbf{0}, 1)$. Using Eq.(2.9) we get

$$F(\mathbf{0}, z) = 1 - \frac{1}{P(\mathbf{0}, z)}, \tag{2.11}$$

and respectively

$$F(\mathbf{0}) = 1 - \frac{1}{P(\mathbf{0}, 1)}.$$

From knowledge of $P(\mathbf{0}, 1)$ we can obtain quite easily the probability of return. In the following we give examples of how to calculate this return probability explicitly.

2.3.2 One-dimensional walk

For a symmetric one-dimensional ($d = 1$) random walk ($\lambda(\theta) = \cos\theta$) the characteristic function of the distribution of the walker's position after n steps is given by $P_n(\theta) = \lambda^n(\theta)$, and the probability of finding a walker at a point x is given by

$$P_n(x) = \frac{1}{2\pi} \int_{-\pi}^{\pi} \lambda^n(\theta) e^{i\theta x}\, d\theta.$$

This means that

$$\begin{aligned} P(0, z) &= \sum_{n=0}^{\infty} P_n(0) z^n = \sum_{n=0}^{\infty} \frac{1}{2\pi} \int_{-\pi}^{\pi} \lambda^n(\theta) z^n \, d\theta \\ &= \frac{1}{2\pi} \int_{-\pi}^{\pi} \left[\sum_{n=0}^{\infty} \lambda^n(\theta) z^n \right] d\theta = \frac{1}{2\pi} \int_{-\pi}^{\pi} \frac{d\theta}{1 - z\lambda(\theta)}, \end{aligned} \tag{2.12}$$

where in the second line we simply summed up the geometrical series. In our case Eq.(2.12) reads:

$$P(0, z) = \frac{1}{2\pi} \int_{-\pi}^{\pi} \frac{d\theta}{1 - z\cos\theta}.$$

The corresponding integral is known (Eq.(3.613.1) of Ref.[6]). For $-1 < z < 1$ we have

$$\frac{1}{2\pi} \int_{-\pi}^{\pi} \frac{d\theta}{1 - z\cos\theta} = \frac{1}{\sqrt{1 - z^2}}.$$

According to Eq.(2.11) we have

$$F(0, z) = 1 - \sqrt{1 - z^2}. \tag{2.13}$$

We immediately see that $F(0) = F(0, 1) = 1$, so that the one-dimensional symmetric random walk is recurrent.

Since the Taylor expansion of the square root is known, it is possible to show that

$$F_n(0) = \begin{cases} \frac{2}{n-1} \binom{n-1}{n/2} 2^{-n} & n \text{ even} \\ 0 & n \text{ odd}, \end{cases} \tag{2.14}$$

since all odd Taylor coefficients for an even function $F(0,\ z)$ vanish. This represents the fact that returns are possible only at even steps.

Exercise 2.6 Show that applying Stirling's formula $\ln n! \approx n \ln n - n + \ln \sqrt{2\pi n}$ to Eq.(2.14) leads, for large n, to

$$F_n(0) = \sqrt{\frac{2}{\pi}} n^{-3/2}, \tag{2.15}$$

which follows a discrete power law. Show that the mean value of n diverges.

We note that, even if an exact form such as Eq.(2.14) is not known, or is too difficult to obtain, the asymptotic result, Eq.(2.15), could still be easily obtained by use of the Tauberian theorem, as in Example 2.2. The result, Eq.(2.15), is much more general than it might seem, and relies essentially only on the symmetry of the steps of the walker, as we show in Sec. 2.5.

Note that in general we have to distinguish between the first return to the origin and the first passage through it. For random walks not restricted to the nearest neighbors on the lattice, or for off-lattice walks, the first passage through the origin has to be understood as follows: The walker starting at the origin 0 takes its first step either in the positive or in the negative direction. In the first case the first passage through the origin corresponds to step n when the coordinate x_n of the walker becomes non-positive for the first time, while in the second case the first passage through the origin corresponds to such an n that x_n becomes non-negative. For nearest-neighbor lattice random walks, the first passage through and the first return to the origin coincide. For any symmetric random walk (independent of the existence of moments) the first-passage probability through the origin behaves for large n asymptotically as

$$F_n(0) = \frac{1}{2\sqrt{\pi}} n^{-3/2}. \tag{2.16}$$

The result, Eq.(2.16), is known as the Sparre Andersen theorem (formulated in 1953) [7], and is discussed in Sec. 2.5. The difference in the prefactor compared to Eq.(2.15) has to do with the fact that, contrary to the nearest-neighbor walks, the first passage through the origin can now also take place at odd steps.

2.3.3 Random walks in higher dimensions

Equation (2.11) is quite general and holds for any dimension. It is clear from this equation that a random walker returns to the origin with unit probability whenever $P(\mathbf{0},1)$ diverges, and the return probability remains less than unity whenever $P(\mathbf{0},1)$ is finite. For example, in the one-dimensional case

$$P(0,z) = \frac{1}{\sqrt{1-z^2}}, \tag{2.17}$$

which diverges for $z \to 1$. Therefore, symmetric one-dimensional walks are recurrent, as explicitly shown above.

Just as in the one-dimensional case, in higher dimensions

$$P_n(\mathbf{0}) = \left(\frac{1}{2\pi}\right)^d \int_\Omega \lambda^n(\mathbf{\Theta}) d\mathbf{\Theta},$$

where the integration takes place over the elementary cell of a lattice. Here we first limit ourselves to a square lattice in $d = 2$ or to a cubic lattice in $d = 3$. Performing the same mathematical steps as above we end with

$$P(\mathbf{0},z) = \left(\frac{1}{2\pi}\right)^d \int_\Omega \frac{d\mathbf{\Theta}}{1 - z\lambda(\mathbf{\Theta})}. \tag{2.18}$$

The integral for $P(\mathbf{0},1)$ diverges only if for some $\mathbf{\Theta}$-vectors $\lambda(\mathbf{\Theta}) = 1$. For example, for a square or a cubic lattice with $\lambda(\mathbf{\Theta}) = \frac{1}{d}\sum_{j=1}^d \cos\theta_j$ the divergence takes place only for $\mathbf{\Theta} = 0$. As discussed in Chapter 1, for small $\mathbf{\Theta}$, $\lambda(\mathbf{\Theta}) \approx 1 - \frac{1}{2d}\theta^2 + \ldots$, so that changing to polar (or spherical) coordinates, we get

$$P(\mathbf{0},1) \approx \left(\frac{1}{2\pi}\right)^d \int_\Omega \frac{S(d)\theta^{d-1}d\theta}{1-(1-\theta^2/2d)} \approx \frac{2dS(d)}{(2\pi)^d} \int_\Omega \theta^{d-3} d\theta,$$

with $S(2) = 2\pi$ and $S(3) = 4\pi$. The corresponding integral diverges for $d = 2$, but converges for $d = 3$.

This means that in two dimensions, just as in one dimension, the symmetric random walks are recurrent, which is a famous result found by Polya (1921) [8]. On the other hand, in three dimensions the random walks are transient. The calculation of the return probability in $d = 3$ assumes the accurate evaluation of the corresponding integral, which has been done for many different lattices. Here we reproduce some known results for cubic lattices [9]:

$$F(\mathbf{0}) = \begin{cases} 0.3405 \text{ for simple cubic lattice} \\ 0.2822 \text{ for body-centered cubic lattice} \\ 0.2563 \text{ for face-centered cubic lattice.} \end{cases}$$

Symmetric random walks in one and in two dimensions are recurrent: the walker returns to its initial position with probability 1.

Symmetric walks in three and higher dimensions are transient.

The recurrence in $d = 1$ is somewhat evident from elementary considerations: Non-recurrent are only those realizations in which, for each n, the number of steps made in one direction is larger than the number of steps in the opposite direction, which is rather improbable.

2.4 Mean number of distinct visited sites

Even the non-recurrent walk revisits from time to time the already visited sites in a lattice; a recurrent walk typically visits sites many times. In various applications, for example in chemical kinetics and search problems, each visited site is counted only once, and therefore the relevant quantity is the number of distinct sites visited in n steps, S_n. The situation is illustrated in Fig. 2.2. The most basic characteristic in this case is given by the *mean number of distinct visited sites*, $\langle S_n \rangle$, which is evaluated here as an example of a more advanced use of generating functions.

To calculate $\langle S_n \rangle$ we note that

$$\langle S_n \rangle = 1 + \sum_{j=1}^{n} \Delta_j. \tag{2.19}$$

Here the unity represents the site the random walk originated from, and Δ_j is the mean number of sites visited for the first time at step j. The latter can be obtained by summing up the probabilities of visiting any site of the lattice at step j for the first time:

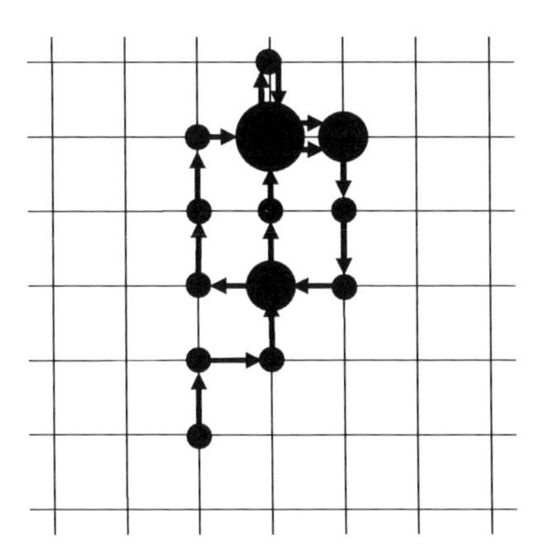

Fig. 2.2 *The random walker in this picture has made* $n = 17$ *steps and visited* $S_n = 14$ *distinct sites. Two sites were visited twice and one site three times.*

$$\Delta_j = \sum_{\mathbf{r}} F_j(\mathbf{r}). \tag{2.20}$$

Let us proceed with evaluating these quantities. We introduce now a new generating function $\Delta(z) = \sum_j \Delta_j z^j$ and express it through the generating function $F(\mathbf{r}, z)$, which itself is expressed in terms of $P(\mathbf{r}, z)$ according to Eq.(2.9): $F(\mathbf{r}, z) = P(\mathbf{r}, z)/P(\mathbf{0}, z) - \delta_{\mathbf{r},0}/P(\mathbf{0}, z)$.

Exercise 2.7 Show that $\sum_{\mathbf{r}} P(\mathbf{r}, z) = \frac{1}{1-z}$.

The generating function $\Delta(z)$ is thus

$$\Delta(z) = \frac{z}{(1-z)P(\mathbf{0}, z)},$$

from which we obtain the generating function $S(z)$ of $\langle S_n \rangle - 1 = \sum_{j=1}^{n} \Delta_j$ as

$$S(z) = \frac{z}{(1-z)^2 P(\mathbf{0}, z)}. \tag{2.21}$$

To verify Eq.(2.21) it is enough to note that $\sum_{j=1}^{n} \Delta_j$ can be expressed in the form $\sum_{j=1}^{n} 1 \cdot \Delta_j$, which is a convolution of a sequence with elements $g_N = 1$ and of the $\{\Delta_j\}$ sequence. As shown earlier, the generating function of the first sequence is $1/(1-z)$.

- For $d = 1$ the expression for $P(0, z)$ is given by Eq.(2.17) so that

$$S(z) = \frac{z\sqrt{1-z^2}}{(1-z)^2} \approx \frac{\sqrt{2}}{(1-z)^{3/2}}.$$

The use of the Tauberian theorem gives us

$$\langle S_n \rangle \approx \frac{\sqrt{2}}{\Gamma(3/2)} n^{1/2} = \sqrt{(8/\pi)n}. \tag{2.22}$$

- In $d = 2$ the integral

$$P(\mathbf{0}, z) = \left(\frac{1}{2\pi}\right)^2 \int_{-\pi}^{\pi} \int_{-\pi}^{\pi} \frac{dk_x dk_y}{1 - z(\cos k_x + \cos k_y)/2}$$

can be expressed in terms of elliptic functions. Its asymptotic behavior for $z \to 1$ is, however, simple:

$$P(\mathbf{0}, z) \cong \frac{1}{\pi} \ln \left(\frac{1}{1-z} \right). \tag{2.23}$$

Using the Tauberian theorem again we get

$$\langle S_n \rangle \cong \pi n / \ln n. \tag{2.24}$$

Exercise 2.8 Prove the asymptotic expression, Eq.(2.23), starting from the integral in polar coordinates

$$P(\mathbf{0}, z) \approx \left(\frac{1}{2\pi} \right)^2 \int_0^{k_{\max}} \frac{2\pi \, kdk}{1 - z(1 - k^2/4)}.$$

- In $d = 3$, $P(\mathbf{0}, z)$ tends to a finite limit for $z \to 1$, so that

$$\langle S_n \rangle \cong n / P(\mathbf{0}, 1). \tag{2.25}$$

Note that in $d = 1$ and $d = 2$ the mean number of distinct visited sites grows more slowly than n, which is a consequence of the "oversampling," because of which each of these sites is visited repeatedly and the number of such visits grows with n. This fact has to do with what is called *compact visitation*, an issue that will be discussed later. On the other hand, in $d = 3$ the number of different visited sites grows as n; the sites are visited only a finite number of times.

The mean number of distinct visited sites $\langle S_n \rangle$ behaves as

$$\langle S_n \rangle \cong \sqrt{(2/\pi) n},$$
$$\langle S_n \rangle \cong \pi n / \ln n,$$

and

$$\langle S_n \rangle \cong n / P(\mathbf{0}, 1)$$

in one, two, and three (and higher) dimensions, respectively.

2.5 Sparre Andersen theorem[2]

Up to now we have discussed the first-passage properties of lattice random walks and stated that Eq.(2.15) for a one-dimensional random walk can be generalized to a case of off-lattice walks with a continuous jump length PDF $p(x)$. Here we consider

[2] This section can be omitted at first reading. The initial derivation, Ref.[8], differs from our discussion here, which follows the lines of Chap. XVIII of Ref.[4].

this continuous case, which brings some surprises. The particle starts at $x = 0$, and we are looking for the probability F_n^+ of first landing on the positive half-line $x > 0$ after n steps. At variance with the discrete situation, we have to keep in mind that the *passing* from one half-line to another half-line does not necessarily correspond to *hitting* a boundary point 0.

Let us introduce two functions, $p_n^+(x)$ and $p_n^-(x)$, which have the following meanings: $p_n^+(x)dx$ gives the probability that a random walker that just crossed from $x \leq 0$ to $x > 0$ at step n for the first time (and never visited the positive half-line $x > 0$ previously) landed in an interval between x and $x + dx$. The function $p_n^-(x)dx$ gives the probability of finding a walker that never visited $x > 0$ up to step n in the interval $(x,\ x + dx)$. The sum $p_n^+(x) + p_n^-(x)$ is a probability density of finding a particle that never visited $x > 0$ up to the step n at some position x on the whole line and is a proper PDF (i.e., is normalized). The sequence of $F_n^+ = \int_0^\infty p_n^+(x)dx$ gives us exactly the first-passage probabilities we are looking for.

To proceed, we first note that $p_n^+(x) = 0$ everywhere on the non-positive half-axis $x \leq 0$ and that $p_n^-(x) = 0$ for any $x > 0$. These key properties of the functions $p_n^\pm(x)$ will be repeatedly used in what follows. From these properties it follows that the convolutions of different $p_m^-(x)$-functions like $\int p_m^-(x')p_n^-(x - x')dx'$ vanish identically for all $x > 0$, and the corresponding convolutions of $p_n^+(x)$-functions vanish for $x \leq 0$. The same is valid for multiple convolutions of more than two functions of the same type.

We now introduce the Fourier transforms

$$p_n^-(k) = \int_{-\infty}^{0} p_n^-(x)\, e^{ikx} dx$$

$$p_n^+(k) = \int_{0+}^{\infty} p_n^+(x)\, e^{ikx} dx,$$

where the limits of integration simply denote the domains where the corresponding functions do not vanish, and the corresponding generating functions

$$\phi^-(z, k) = \sum_{n=0}^{\infty} z^n p_n^-(k) \tag{2.26}$$

and

$$\phi^+(z, k) = \sum_{n=0}^{\infty} z^n p_n^+(k). \tag{2.27}$$

We note that the inverse Fourier transforms of the ϕ-functions

$$\frac{1}{2\pi} \int_{-\infty}^{\infty} \phi^-(z, k)\, e^{-ikx} dk = \sum_{n=0}^{\infty} z^n p_n^-(x)$$

and

$$\frac{1}{2\pi}\int_{-\infty}^{\infty}\phi^{+}(z,k)\,e^{-ikx}dk = \sum_{n=0}^{\infty} z^n p_n^{+}(x)$$

vanish identically for $x > 0$ and for $x \leq 0$ respectively for any value of z. The same is true for the inverse Fourier transforms of the powers of the corresponding ϕ-functions that represent convolutions. Note also that the generating function of the first-passage probabilities $\Phi^{+}(z) = \sum_{n=0}^{\infty} z^n P_n^{+}$ is simply given by

$$\Phi^{+}(z) = \sum_{n=0}^{\infty} z^n \int_{0+}^{\infty} p_n^{+}(x)dx = \phi^{+}(z,0). \tag{2.28}$$

This is the function we are looking for.

The initial conditions at the zeroth step are given by $p_0^{-}(x) = \delta(x)$ and $p_0^{+}(x) = 0$. The last means that the summation in the expression for $\phi^{+}(z,k)$ starts essentially from $n = 1$. Now it is possible to obtain the recursive relation between $p_n^{+}(x)$ and $p_n^{-}(x)$ with different n: If the walker has not visited the half-line $x > 0$ up to the step n and its position at the n-th step is y, then at step $n+1$ it may either stay on the half-line $x \leq 0$ or jump over 0 and cross to the positive half-line. Therefore it holds:

$$p_{n+1}^{-}(x) + p_{n+1}^{+}(x) = \int_{-\infty}^{0} p_n^{-}(y)p(x-y)dy.$$

Applying the Fourier transform, we obtain

$$p_{n+1}^{-}(k) + p_{n+1}^{+}(k) = p_n^{-}(k)\lambda(k). \tag{2.29}$$

This expression can be used to build up the generating functions (2.26) and (2.27). Multiplying equations (2.29) for different n by z^{n+1} and summing up the results, we get:

$$\sum_{n=0}^{\infty} z^{n+1}p_{n+1}^{-}(k) + \sum_{n=0}^{\infty} z^{n+1}p_{n+1}^{+}(k) = \lambda(k)\sum_{n=0}^{\infty} z^{n+1}p_n^{-}(k).$$

On the right-hand side we rename the summation index to $l = n+1$ and note that the first sum on the left-hand side is $\phi^{-}(z,k) - 1$ (since the zero-order term, being the Fourier transform of the initial condition in the form of the δ-function, does not appear in the sum), and that the second sum is exactly $\phi^{+}(z,k)$. Thus, we get $\phi^{+}(z,k) + \phi^{-}(z,k) - 1 = z\lambda(k)\phi^{-}(z,k)$, which we rewrite as

$$1 - \phi^{+}(z,k) = \phi^{-}(z,k)\left[1 - z\lambda(k)\right]. \tag{2.30}$$

Equation (2.30) gives *one* condition for *two* unknown functions $\phi^+(z,k)$ and $\phi^-(z,k)$; however, it is sufficient to uniquely define $\Phi^+(z) = \phi^+(z,0)$, because the functions $p_n^+(x)$ and $p_n^-(x)$ never differ from zero simultaneously.

Assuming that $\phi^+(z,k)$ and $\phi^-(z,k)$ do not have zeroes within their domains of definition (we omit here the proof of this fact), we can take the logarithms of both parts of the equation to obtain

$$\ln\left[1-\phi^+(z,k)\right] = \ln\phi^-(z,k) + \ln\left[1 - z\lambda(k)\right]. \tag{2.31}$$

Now we apply the inverse Fourier transform to both parts of this equation

$$\frac{1}{2\pi}\int_{-\infty}^{\infty} \ln\left[1-\phi^+(z,k)\right] e^{-ikx} dk = \frac{1}{2\pi}\int_{-\infty}^{\infty} \ln\phi^-(z,k) e^{-ikx} dk + \frac{1}{2\pi}\int_{-\infty}^{\infty} \ln\left[1-z\lambda(k)\right] e^{-ikx} dk$$

and take $x > 0$. Let us discuss the behavior of the terms of this equation, starting from the last term on the right-hand side. Expanding the last term of Eq.(2.31) into Taylor series we get

$$\ln\left[1 - z\lambda(k)\right] = -\sum_{n=1}^{\infty}\frac{z^n}{n}\lambda^n(k),$$

so that its inverse Fourier transform equals

$$-\sum_{n=1}^{\infty}\frac{z^n}{n}P_n(x),$$

where $P_n(x)$ are the PDFs of the particle's position at step n. The other term on the right-hand side of Eq.(2.31) has the form $\ln\left[\sum_{n=0}^{\infty} z^n p_n^-(k)\right] = \ln\left[1 + \sum_{n=1}^{\infty} z^n p_n^-(k)\right]$. Expanding it into Taylor series, we see that its n-th term $(n \geq 1)$ has the form of a sum of products of $p_m^-(k)$ of the type $A_1 p_n^-(k) + A_2 p_{n_1}^-(k) p_{n_2}^-(k) + \ldots + A_m p_{n_1}^-(k) p_{n_2}^-(k) \ldots p_{n_m}^-(k) + \ldots$ with $n_1 + n_2 + \ldots = n$, which are the Fourier transforms of the convolutions of different p^--functions. Thus, the inverse Fourier transform of the whole series represents a function that vanishes identically for all $x > 0$. On the other hand, the inverse Fourier transform of the function on the left-hand side of the equation,

$$\ln\left[1-\phi^+(z,k)\right] = -\sum_{n=1}^{\infty}\frac{\left[\phi^+(z,k)\right]^n}{n}, \tag{2.32}$$

represents a function that vanishes identically for all $x \leq 0$. Therefore taking the inverse Fourier transform of both parts of the equation for some $x > 0$ we can neglect the first term on the right-hand side and write

$$\frac{1}{2\pi}\int_{-\infty}^{\infty}\ln\left[1-\phi^{+}(z,k)\right]e^{-ikx}dk=-\sum_{n=1}^{\infty}\frac{z^n}{n}P_n(x)$$

for all $x > 0$. Let us now take the integral over all positive x of both parts of this equation:

$$\int_{0+}^{\infty}\frac{1}{2\pi}\int_{-\infty}^{\infty}\ln\left[1-\phi^{+}(z,k)\right]e^{-ikx}dkdx=-\int_{0+}^{\infty}\sum_{n=1}^{\infty}\frac{z^n}{n}P_n(x)dx. \tag{2.33}$$

Changing the sequence of integration and summation on the right-hand side, we get

$$-\int_{0+}^{\infty}\sum_{n=1}^{\infty}\frac{z^n}{n}p_n(x)dx=-\sum_{n=1}^{\infty}\frac{z^n}{n}C_n(x>0),$$

where $C_n(x>0)=\int_{0+}^{\infty}P_n(x)dx$ is the cumulative probability of finding a walker on the positive half-axis after n steps.

Let us turn to the left-hand side of Eq.(2.33). Noting that $\ln\left[1-\phi^{+}(z,k)\right]$ is a Fourier transform of a function that vanishes identically for $x \leq 0$, as it follows from Eq.(2.32), the integration over x can be extended to the whole real axis. Then the sequence of integrations over k and over x can be reversed. It follows that the expression on the left-hand side of Eq.(2.33) corresponds to taking the limit $k \to 0$ of the integrand. Therefore

$$\ln\left[1-\phi^{+}(z,0)\right]=-\sum_{n=1}^{\infty}\frac{z^n}{n}C_n(x>0).$$

Recalling that according to Eq.(2.28) the function $\Phi^{+}(z)=\phi^{+}(z,0)$ is the generating function of the first-passage probabilities to $x > 0$, we obtain the main result of the Sparre Andersen theorem, namely that

$$\Phi^{+}(z)=1-\exp\left[-\sum_{n=1}^{\infty}\frac{z^n}{n}C_n(x>0)\right]. \tag{2.34}$$

Let us now consider the case of a *symmetric* random walk starting at $x = 0$ and look at the probabilities of first passage to the right half-line $x > 0$. Owing to the symmetry of the walk, the probability of finding a walker on a positive half-line after any step is $C_n(x>0)=1/2$, so that we obtain:

$$\Phi^{+}(z)=1-\exp\left[-\frac{1}{2}\sum_{n=1}^{\infty}\frac{z^n}{n}\right]=1-\exp\left[\frac{1}{2}\ln(1-z)\right]=1-\sqrt{1-z}. \tag{2.35}$$

This is exactly the generating function of Exercise 2.4. For those who did not consider the exercise, we show another way to obtain the first-passage probabilities. Using the binomial expansion for a square root, we get

$$1-\sqrt{1-z}=1-\sum_{n=0}^{\infty}\binom{1/2}{n}(-z)^n \equiv \sum_{n=0}^{\infty}\frac{\left(\frac{1}{2}\right)\left(\frac{1}{2}-1\right)\ldots\left(\frac{1}{2}-n+1\right)}{n!}(-z)^n$$

$$=\sum_{n=0}^{\infty}\frac{\Gamma\left(n-\frac{1}{2}\right)}{\Gamma\left(\frac{1}{2}\right)}\frac{z^n}{n!}.$$

We now use Stirling's formula to obtain the asymptotic form of the coefficients:

$$\ln\frac{\Gamma\left(n-\frac{1}{2}\right)}{2n!\Gamma\left(\frac{1}{2}\right)} \approx \ln\frac{1}{2\sqrt{\pi}}-\frac{3}{2}\ln n$$

so that

$$F_N^+ \cong \frac{1}{2\sqrt{\pi}}n^{-3/2}. \tag{2.36}$$

Note also that this result is practically the same as for a simple discrete random walk (the difference in the prefactor has to do with the fact that in a simple discrete walk the returns to the origin can take place only at even steps)!

Our derivation is based on adapting the mathematical methods discussed in Chapter XVIII of Ref.[4] to the level of knowledge the reader is assumed to gain from Chapters 1 and 2 of the present book. The advantage is that it closely follows the methods we have applied dealing with random walks on lattices and uses only the means introduced in Chapters 1 and 2. We have sacrificed a great deal of mathematical rigor in favor of simplicity and transparency.

Independent of the exact form of the PDF of the displacement per step and even of whether the moments of this displacement exist, the first-passage probability to the positive half-line of a symmetric random walk behaves as

$$F_n^+ \cong \frac{1}{2\sqrt{\pi}}n^{-3/2}$$

(as a consequence of the Sparre Andersen theorem).

References

[1] G. Doetsch. *Introduction to the Theory and Application of the Laplace Transformation*, Berlin: Springer, 1974

[2] F. Oberhettinger. *Tables of Laplace Transforms*, New York: Springer, 1973

[3] A.P. Prudnikov, Ya.A. Brychkov, and O.I. Marichev. *Integrals and Series*, Vol. 4: *Direct Laplace Transforms*; Vol. 5: *Inverse Laplace Transforms*, New York: Gordon and Breach, 1992

[4] W. Feller. *An Introduction to Probability Theory and Its Applications*, New York: Wiley, 1971 (Tauberian theorems are discussed in Vol. 2 Chap. XIII, Sec. 2.5).
[5] M. Abramovitz and I.A. Stegun. *Handbook of Mathematical Functions*, New York: Dover, 1972
[6] I.S. Gradsteyn and I.M. Ryzhik. *Table of Integrals, Series, and Products*, San Diego: Academic Press, 1980
[7] E. Sparre Andersen. *Math. Scand.* **1**, 263 (1953)
[8] G. Polya. *Math. Ann.* **84**, 149 (1921)
[9] E.W. Montroll and G.H. Weiss. *J. Math. Phys.* **6**, 167 (1965)

Further reading

M.N. Barber and B.W. Ninham. *Random and Restricted Walks: Theory and Applications*, New York: Gordon and Breach, 1970
G.H. Weiss. *Aspects and Applications of the Random Walk*, Amsterdam: North-Holland, 1994
B.D. Hughes. *Random Walks and Random Environments*, Vol. 1: *Random Walks*, Oxford: Clarendon, 1996
S. Redner. *A Guide to First-Passage Processes*, Cambridge: Cambridge University Press, 2001

3

Continuous-time random walks

"Everything happens to everybody sooner or later if it is time enough."

George Bernard Shaw

In our world, we are interested in processes evolving in continuous physical time. In Chapter 1, however, we introduced random walks parameterized by the number of steps, which can be considered as the internal discrete time of the system. The aim of the present chapter is to translate the number of steps into real time. Physical time is introduced here in terms of continuous-time random walks (CTRWs) as formulated by E. Montroll and G.H. Weiss (1964) [1].

3.1 Waiting-time distributions

In CTRW a random walker jumps instantaneously from one site to another, following a waiting period on a site whose duration t is drawn according to a PDF of waiting times $\psi(t)$ (see Fig. 3.1).

The main input is therefore the form of this probability density. If all waiting times are equal, $\psi(t) = \delta(t - \tau)$, each step of a random walk has the same time cost τ, which leads to a simple way to translate the number of steps n into time t. This situation, however, is an exception; any disorder in waiting times leads to a more complicated process. This process will be considered now under the assumption that the waiting times of subsequent steps (namely t_1, $t_2, \ldots$ in Fig. 3.1) are independent. A trajectory of a one-dimensional CTRW with an exponential waiting-time PDF $\psi(t) = \lambda e^{-\lambda t}$ is shown in Fig. 3.2.

Our discussion here will be very similar to that in Chapter 1. However, the fact that the waiting times are non-negative makes it reasonable to use another mathematical

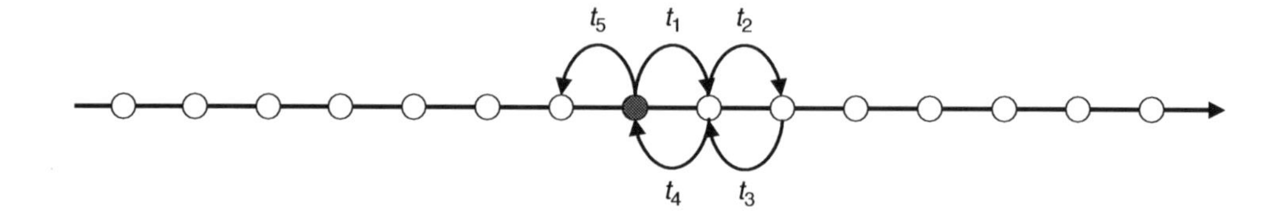

Fig. 3.1 *A schematic look at CTRWs.*

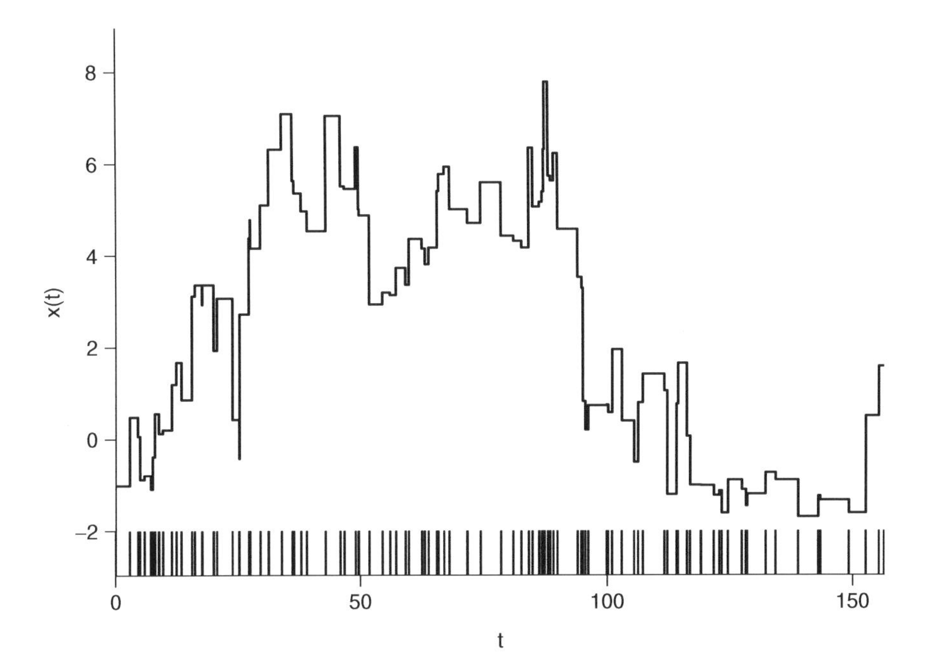

Fig. 3.2 *A trajectory of a CTRW with exponential distribution of waiting times* ($\lambda = 1$) *and Gaussian distribution of step lengths (with zero mean and unit dispersion). The step times are shown as the barcode-like pattern in the lower part of the graph.*

instrument, namely the Laplace transform [2] (see also Ref.[3]) (instead of the Fourier transform), of the probability density,

$$\psi(s) = \int_0^\infty e^{-st}\psi(t)dt \equiv \left\langle e^{-st} \right\rangle . \tag{3.1}$$

Note that since the PDF of the waiting times $\psi(t)$ is normalized, $\int_0^\infty \psi(t)dt = 1$, its Laplace transform satisfies the condition $\psi(s \to 0) = 1$. Just like the characteristic function (Fourier transform), the Laplace transform of $\psi(t)$ is the generating function of its moments provided they exist:

$$\psi(s) = \sum_{n=0}^\infty (-1)^n \frac{\langle t^n \rangle s^n}{n!}. \tag{3.2}$$

Exercise 3.1 Consider the exponential waiting-time PDF $\psi(t) = \lambda \exp(-\lambda t)$. Calculate the mean waiting time and the variance of the waiting times. This can be done using both the definition of moments and the generating function, Eq.(3.2).

Exercise 3.2 Now we consider the stretched-exponential waiting-time PDF $\psi(t) = f(b,\alpha)\exp(-bt^{\alpha})$. Find the normalization factor $f(b,\alpha)$. Find the mean waiting time corresponding to such a distribution.

Hint: Use the change of variables $\xi = bt^{\alpha}$ and the definition of Γ-function (see Ref.[2]).

Let us consider a process starting at $t = 0$ and define $\psi_n(t)$, which is the probability density of the occurrence of the n-th step at time $t = t_1 + t_2 + \ldots + t_n$, where t_i is the waiting time for the i-th step of the walker (see Fig. 3.1). In the Laplace domain $\psi_n(t)$ corresponds to

$$\psi_n(s) = \left\langle e^{-s(t_1+t_2+\ldots+t_n)} \right\rangle = \left\langle e^{-st_1} e^{-st_2} \ldots e^{-st_n} \right\rangle = \left\langle e^{-st_1} \right\rangle \left\langle e^{-st_2} \right\rangle \ldots \left\langle e^{-st_n} \right\rangle$$
$$= \left\langle e^{-st} \right\rangle^n = \psi^n(s), \tag{3.3}$$

where, just as in Sec. 1.4, the possibility of writing a mean value of a product as a product of mean values stems from the independence of the multipliers. There is a full similarity between this equation and Eq.(1.5) of Chapter 1. The analog of Eq.(1.4) reads

$$\psi_n(t) = \int_0^t \psi_{n-1}(t')\psi(t-t')dt'.$$

The Laplace transform of this convolution leads to $\psi_n(s) = \psi_{n-1}(s)\psi(s)$, which, after being iterated, leads to Eq.(3.3).

Another important quantity is the so-called survival probability on a site, the probability that the waiting time on a site exceeds t:

$$\Psi(t) = \int_t^{\infty} \psi(t')dt' = 1 - \int_0^t \psi(t')dt'. \tag{3.4}$$

The Laplace transform of $\Psi(t)$ follows from the form for the Laplace transform of an integral [2] and reads

$$\Psi(s) = \frac{1}{s} - \frac{\psi(s)}{s} = \frac{1-\psi(s)}{s}. \tag{3.5}$$

Introducing this quantity allows us to obtain the probability of taking exactly n steps up to the time t

$$\chi_n(t) = \int_0^t \psi_n(\tau)\Psi(t-\tau)d\tau, \tag{3.6}$$

which corresponds to completing n steps at time $\tau < t$ and waiting at rest for the remainder of the time, $t - \tau$. Note the difference between the *probability* $\chi_n(t)$ that up to time t exactly n steps are taken and the *probability density* $\psi_n(t)$ that the n-th step follows at time t. The connection between the two is given by Eq.(3.6). Since this equation also has the form of a convolution, its Laplace transform reads

$$\chi_n(s) = \psi^n(s)\frac{1-\psi(s)}{s}. \tag{3.7}$$

The correspondence of the internal clock n of a CTRW and the physical time t is given by $\chi_n(t)$, the probability of taking exactly n steps up to time t. Its Laplace transform reads:

$$\chi_n(s) = \psi^n(s)\frac{1-\psi(s)}{s}.$$

3.2 Transforming steps into time

We can now return to our random walk and note that the PDF of the position of a walker at time t is given by

$$P(x,t) = \sum_{n=0}^{\infty} P_n(x)\chi_n(t), \tag{3.8}$$

which translates the process of Chapter 1 from its internal clock n to the physical time t we are interested in. Equation (3.8) has a simple meaning: The position of a random walker at time t is its position after n steps, provided exactly n steps were made up to time t, which happens with probability $\chi_n(t)$. Equation (3.8) is essentially an example of *subordination*, discussed at the end of this chapter. Now we Laplace transform Eq.(3.8) and obtain

$$P(x,s) = \sum_{n=0}^{\infty} P_n(x)\chi_n(s) = \sum_{n=0}^{\infty} P_n(x)\psi^n(s)\frac{1-\psi(s)}{s}. \tag{3.9}$$

Fourier transform from x to k and using Eq.(1.5) in Chapter 1 leads to the following expression for $P(k,s)$:

$$P(k,s) = \sum_{n=0}^{\infty} P_n(k)\chi_n(s) = \frac{1-\psi(s)}{s}\sum_{n=0}^{\infty}\lambda^n(k)\psi^n(s).$$

The sum on the right-hand side of this equation is a geometric series, which can be easily summed up leading to a closed expression for $P(k,s)$:

$$P(k,s) = \frac{1-\psi(s)}{s}\frac{1}{1-\lambda(k)\psi(s)}. \tag{3.10}$$

This equation is the central result of the theory of CTRWs (at least of decoupled walks). The inverse transform (Laplace and Fourier) gives $P(x,t)$, a PDF to find a walker at position x at time t. We note that Eq.(1.5) in Chapter 1 implies that the random walker starts at $x=0$ before performing its first step, and therefore Eq.(3.10) applies to a situation when a walker starts at $x=0$ at time $t=0$. The relatively simple expression Eq.(3.10) shows again that it is easier to work in the Fourier and Laplace domains than in real space and time. Equation (3.10) retains its form also in higher dimensions.

The main result for the decoupled CTRWs is given by the Fourier–Laplace transform of the PDF to find a walker at position $\mathbf{r}$ at time t:

$$P(\mathbf{k},s)=\frac{1-\psi(s)}{s}\frac{1}{1-\lambda(\mathbf{k})\psi(s)}.$$

Example 3.1 Let us assume the waiting-time PDF to be exponential,

$$\psi(t)=\tau^{-1}e^{-t/\tau}$$

so that its Laplace transform is

$$\psi(s)=\frac{1}{1+s\tau}. \tag{3.11}$$

Parallel to the case of Fourier transform, small values of s correspond to long times. As follows from Eq.(3.2) for small s, $\psi(s)$ can be approximated by $\psi(s)=1-\langle t\rangle s+\ldots$, with $\langle t\rangle=\tau$. Let us moreover assume the PDF of step displacements to be Gaussian, as in Example 1.1 in Chapter 1. Substituting the corresponding $\lambda(k)$ and $\psi(s)$ into Eq.(3.10) and expanding both $\lambda(k)$ and $\psi(s)$ up to the terms of the order of k^2 and s^1 we get

$$P(k,s)\approx\frac{\tau}{1-\left(1-\frac{k^2\sigma^2}{2}\right)(1-s\tau)}.$$

Neglecting the mixed terms of the higher order (i.e., the term containing k^2s) we obtain

$$P(k,s)\approx\frac{\tau}{\frac{k^2\sigma^2}{2}+s\tau}. \tag{3.12}$$

The inverse transform of this expression to the space–time domain will be done in two steps: First the inverse Laplace transform, which gives

$$P(k,t)=\exp\left(-\frac{\sigma^2k^2}{2\tau}t\right)$$

(since we recognize in Eq.(3.12) a Laplace transform of an exponential function, similar to Eq.(3.11)), and its inverse Fourier transform gives the Gaussian function

$$P(x,t) = \frac{1}{\sqrt{2\pi\sigma^2 t/\tau}} \exp\left(-\frac{x^2}{2\sigma^2 t/\tau}\right).$$

Example 3.2 While the previous situation with the Gaussian distribution of step lengths could be considered only approximately, the case of CTRWs (with exponentially distributed waiting times) on a lattice leads to an exact expression. Here we combine $\psi(s) = 1/(1+s\tau)$ with $\lambda(k) = \cos k$ (which, for the discrete random walks is defined on an interval $-\pi \leq k \leq \pi$) to get

$$P(k,s) = \frac{1 - \frac{1}{1+s\tau}}{s} \frac{1}{1 - \frac{1}{1+s\tau}\cos k} = \frac{\tau}{s\tau + (1-\cos k)}.$$

The inverse Laplace transform of this expression again gives us an exponential function, now

$$P(k,t) = e^{-(1-\cos k)t/\tau},$$

and its subsequent inverse Fourier transform gives

$$P(j,t) = \frac{1}{2\pi} \int_{-\pi}^{\pi} \cos jk \; e^{-(1-\cos k)t/\tau} dk = e^{-t/\tau} I_j(t/\tau),$$

where $I_j(x)$ is the modified Bessel function (see Eq.(9.6.20) of Ref.[4]). Using the asymptotic expansion Eq.(9.7.1) of Ref.[4],

$$I_\nu(z) \sim \frac{e^z}{\sqrt{2\pi z}} \left\{1 - \frac{4\nu^2 - 1}{8z} + \dots\right\},$$

we see that, e.g., the probability of being at the origin

$$P(0,t) = e^{-t/\tau} I_0(t/\tau) \cong \frac{1}{\sqrt{2\pi t/\tau}}$$

decays as a square root of time. This one is different from the probability of *returning* to the origin, which is considered at the end of this chapter.

Up to now we considered one-dimensional CTRW. Let us now consider the case of higher dimensions. There exists a deep connection between the CTRW on lattices and the generating functions of the lattice random walks discussed in Chapter 2. Let us return to Eq.(3.8):

$$P(\mathbf{r},s) = \sum_{n=0}^{\infty} P_n(\mathbf{r})\chi_n(s) = \frac{1-\psi(s)}{s} \sum_{n=0}^{\infty} P_n(\mathbf{r})\psi^n(s).$$

The sum in this equation corresponds to the generating function of $P(\mathbf{r},z)$, $P(\mathbf{r},z) = \sum_{n=0}^{\infty} P_n(\mathbf{r}) z^n$, with the variable z now replaced for $\psi(s)$. This allows the use of many expressions from Chapter 2 by formally replacing of a variable and multiplying the result by $(1-\psi(s))/s$:

$$P(\mathbf{r},s)=\frac{1-\psi(s)}{s}P(\mathbf{r},z=\psi(s)). \tag{3.13}$$

From the result of Exercise 2.5,

$$P(\mathbf{r},z)=\sum_{n=0}^{\infty}z^nP_n(\mathbf{r})=\frac{1}{(2\pi)^d}\int_{-\pi}^{\pi}\cdots\int_{-\pi}^{\pi}\frac{e^{-i\boldsymbol{\Theta}\mathbf{r}}}{1-zp(\boldsymbol{\Theta})}d^d\boldsymbol{\Theta},$$

we obtain for a CTRW on a discrete d-dimensional hypercubic lattice

$$P(\mathbf{r},s)=\frac{1-\psi(s)}{(2\pi)^d s}\int_{-\pi}^{\pi}\cdots\int_{-\pi}^{\pi}\frac{e^{-i\boldsymbol{\Theta}\mathbf{r}}}{1-\lambda(\boldsymbol{\Theta})\psi(s)}d^d\boldsymbol{\Theta}.$$

This result is explicitly used in Chapter 4.

3.3 Moments of displacement in CTRW

Let us consider the time dependence of the moments of the displacement, $M_n(t)=(-i)^n\left.\frac{d^nP(k,t)}{dk^n}\right|_{k=0}$ (see Chapter 1), which in the Laplace domain leads to

$$M_n(s)=\int_0^{\infty}M_n(t)e^{-st}dt=(-i)^n\left.\frac{d^nP(k,s)}{dk^n}\right|_{k=0}.$$

For example, let us calculate the MSD of a CTRW, a quantity that appears in a plentitude of applications:

$$\begin{aligned}M_2(s)&=-\left.\frac{d^2P(k,s)}{dk^2}\right|_{k=0}=-\frac{1-\psi(s)}{s}\frac{d^2}{dk^2}\left.\frac{1}{1-\lambda(k)\psi(s)}\right|_{k=0}\\&=\frac{\psi(s)}{s\left[1-\psi(s)\right]}\left\langle l^2\right\rangle+\frac{2\psi^2(s)}{s\left[1-\psi(s)\right]^2}\left\langle l\right\rangle^2.\end{aligned}$$

Here we have used the fact that $\lambda(0)=1$ and introduced the notation $\langle l\rangle=\int_{-\infty}^{\infty}xp(x)dx$ and $\langle l^2\rangle=\int_{-\infty}^{\infty}x^2p(x)dx$ for the first two moments of the displacement in one step. For the symmetric CTRW the second term vanishes, so that

$$M_2(s)=\left\langle x^2(s)\right\rangle=\frac{\psi(s)}{s\left[1-\psi(s)\right]}\left\langle l^2\right\rangle. \tag{3.14}$$

Example 3.3 For an exponential waiting-time distribution whose Laplace transform is given by Eq.(3.11), the MSD is

$$\langle x^2(s) \rangle = \frac{\psi(s)}{s\,[1-\psi(s)]} \langle l^2 \rangle = \langle l^2 \rangle \frac{\frac{1}{1+s\tau}}{s\left[1-\frac{1}{1+s\tau}\right]} = \langle l^2 \rangle \frac{1}{s^2\tau}.$$

The inverse Laplace transform is trivial and gives

$$\langle x^2(t) \rangle = \frac{\langle l^2 \rangle}{\tau} t. \tag{3.15}$$

The linear dependence of the MSD on time is quite generic and is always the case (at least asymptotically) if the second moment of a step displacement PDF $p(x)$ and the first moment of waiting-time PDF $\psi(t)$ exist. The prefactor of the time in Eq.(3.15) is connected to the diffusion coefficient D via $D = \langle l^2 \rangle / 2\tau$, since in the normal, Fickian diffusion the mean MSD goes as

$$\langle x^2(t) \rangle = 2Dt \tag{3.16}$$

in one dimension.

Exercise 3.3 Let us consider the mean number of steps done up to time t: $\langle n(t) \rangle = \sum_{n=0}^{\infty} n\chi_n(t)$. In the Laplace domain this expression is given by $\langle n(s) \rangle = \frac{1-\psi(s)}{s} \sum_{n=0}^{\infty} n\psi^n(s)$. Show that

$$\langle n(s) \rangle = \frac{\psi(s)}{s\,[1-\psi(s)]}. \tag{3.17}$$

Hint: $\sum_{n=0}^{\infty} nz^n = z\frac{d}{dz} \sum_{n=0}^{\infty} z^n$.

Note that Eq.(3.17) essentially shows that the MSD given by Eq.(3.14) corresponds to $\langle x^2(s) \rangle = \langle l^2 \rangle \langle n(s) \rangle$ or

$$\langle x^2(t) \rangle = \langle l^2 \rangle \langle n(t) \rangle \tag{3.18}$$

in the Laplace or time domain, respectively. Equation (3.18) is extremely transparent and has a larger domain of applicability than may be apparent from our previous derivation. It corresponds to the average of Eq.(1.11) in Chapter 1 over possible realizations of n.

3.4 Power-law waiting-time distributions

Until now we have concentrated on the waiting-time distributions possessing the first moment, namely $\langle t \rangle = \tau$. There are, however, situations where the waiting-time distributions do not possess moments, in particular the first moment. The most important ones are those that behave *asymptotically* as power laws,

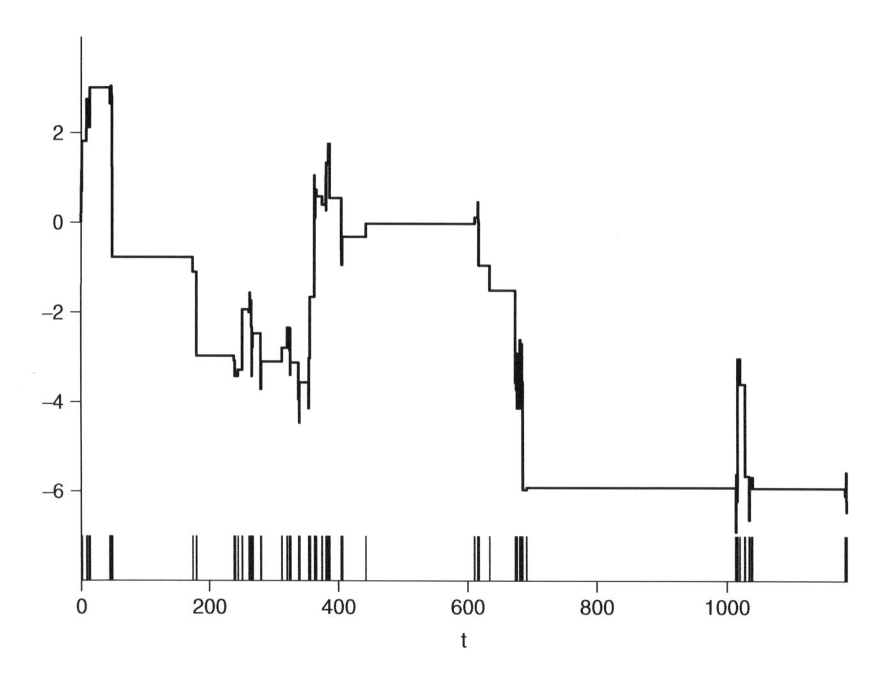

Fig. 3.3 *A realization of the CTRW of 100 steps with $\psi(t) \propto t^{-3/2}$. The step times are shown in the lower part of the graph, as was done in Fig. 3.2. The distribution of step lengths is a symmetric Gaussian with unit dispersion.*

$$\psi(t) \sim \frac{\alpha}{\Gamma(1-\alpha)} \frac{\tau^{\alpha}}{t^{1+\alpha}}, \quad 0 < \alpha < 1, \tag{3.19}$$

also referred to a heavy-tailed waiting-time distributions. The prefactor containing the Gamma function in Eq.(3.19) is introduced in order to simplify some of the further expressions.

A realization of CTRW with power-law waiting-time density $\psi(t) = \frac{1}{6} \frac{1}{(1+t/3)^{3/2}}$ corresponding to $\alpha = 1/2$ is shown in Fig. 3.3. The function $\psi(t)$ is chosen in such a way that its median is exactly unity: With probability 1/2, the waiting time at the site is smaller than unity (and with probability 1/2, it is larger). The barcode-like graph in the bottom of this figure gives the times of jumps in this heavy-tailed CTRW. Note that the jumps follow quite irregularly. As time progresses, the probability of getting stuck for a long time grows. This is in contrast with what is observed in the case of exponential waiting-time distributions, and this is what makes the power-law distributions lacking the first moment so special.

The Laplace transform of the distribution, Eq.(3.19), for small values of s follows from the Tauberian theorems for the Laplace transform (which are closely related to those for series).

The corresponding theorem states that, if for $t \to \infty$ the function $f(t)$ behaves as $f(t) \cong t^{\rho-1}L(t)$, with $0 < \rho < \infty$ and $L(t)$ being a slowly varying function of its argument, its Laplace transform is given by $f(s) \cong \Gamma(\rho)s^{-\rho}L(1/s)$, where $\Gamma(\rho)$ is a Gamma function [5].[1]

According to Tauberian theorems

$$f(t) \cong t^{\rho-1}L(t)$$

$$\Updownarrow$$

$$f(s) \cong \Gamma(\rho)s^{-\rho}L(1/s).$$

However, the immediate application of the Tauberian theorems to Eq.(3.19) does not lead to a correct result, since $\psi(t)$ has to satisfy an additional condition, namely, it is normalized, $\int_0^\infty \psi(t)dt = 1$ and therefore $\psi(s \to 0) = 1$. To satisfy this condition, we consider the function $\Psi(t)$, which for large t behaves as $\Psi(t) \cong \tau^\alpha/t^\alpha$. Application of the Tauberian theorems to this function gives: $\Psi(s) \cong \tau^\alpha s^{\alpha-1}$. Using Eq.(3.5) we then obtain for $\psi(s)$

$$\psi(s) = 1 - \tau^\alpha s^\alpha \tag{3.20}$$

for small s.

The strength of the Tauberian theorems is demonstrated by the following exercises, discussing the case of so-called ultra-slow kinetics where the waiting-time PDFs show the slowest possible decay.

Exercise 3.4 Calculate $\psi(s)$ in a CTRW with the waiting-time PDF that asymptotically behaves as $\psi(t) \sim \frac{1}{t\ln^\beta t}(\beta > 1)$. Show that $1 - \psi(s) \propto \frac{1}{(\beta-1)}\frac{1}{\ln^{\beta-1}(1/s)}$.

Exercise 3.5 Show that the PDF of the particle's displacement in the ultra-slow CTRW (Exercise 3.4) follows a two-sided exponential pattern $P(x,t) = \frac{1}{2W(t)}\exp\left(-\frac{|x|}{W(t)}\right)$. Find the temporal dependence of the width $W(t)$ of this distribution.

Hint: Follow the procedure leading to Eq.(3.12) for the normal case and perform first the inverse Laplace transform by noting that the corresponding $P(k,s)$ has the form $\frac{1}{s}L\left(\frac{1}{s}\right)$ with $L(x)$ being a slowly varying function (depending on k).

The behavior of $\psi(s)$ for the case of power-law waiting-time PDFs is discussed in the following two exercises.

[1] Although mostly used exactly in this setting, the Tauberian theorem also applies to the reverse case, connecting the behavior of the original for $t \to 0$ with the behavior of its Laplace transform for $s \to \infty$. In this case the function $L(t)$ should be slowly varying at zero, which corresponds to $L(1/t)$ being a slowly varying function (at infinity).

Exercise 3.6 When not paying attention to the slowly varying function, the result Eq.(3.20) can be obtained by an approach similar to the one used in Chapter 1, Eq.(1.13). Derive Eq.(3.20) starting from $\psi(s) = 1 - (1 - \psi(s)) = 1 - \int_0^\infty (1 - e^{-st})\psi(t)\,dt$.

Exercise 3.7 The expression for the probability of taking exactly n steps up to time t reads in the Laplace domain as $\chi_n(s) = \psi^n(s)\frac{1-\psi(s)}{s}$. Consider heavy-tailed $\psi(t)$, Eq.(3.19), and show that for $s \to 0$ we have $\chi_n(s) \cong \tau^\alpha s^{\alpha-1}\exp(-n\tau^\alpha s^a)$.

The result of Exercise 3.7 connects the asymptotic form of the distribution of the number of steps made up to time t with the *one-sided Lévy laws* $L_\alpha(t)$, the PDFs whose Laplace transforms are $L_\alpha(s) = \exp(-s^\alpha)$.

Exercise 3.8 Show that the asymptotic form of $\chi_n(t)$ is given by $\chi_n(t) \approx \frac{t}{\alpha\tau} n^{-\frac{1}{\alpha}-1} L_\alpha\left(\frac{t}{\tau n^{1/\alpha}}\right)$.

Hint: Note that $\chi_n(s) \cong \frac{1}{s}\frac{d}{dn}\exp(-n\tau^\alpha s^a)$.

As we discuss in detail in Chapter 7, the one-sided Lévy PDF behaves as $L_\alpha(t) \propto t^{-1-\alpha}$ for $t >> 1$ and decays very fast to zero when $t \to 0$. Therefore $\chi_n(t) \propto \left(\frac{t}{\tau}\right)^\alpha$ for $n << (t/\tau)^\alpha$, and tends to zero for $n >> (t/\tau)^\alpha$ [6], so that in many applications it can be approximated by a power law with a cutoff.

Let us give a few examples of heavy-tailed waiting-time distributions. A very prominent one is the Lévy–Smirnov PDF

$$\psi(t) = \frac{\tau^{1/2}}{2\sqrt{\pi}t^{3/2}}\exp\left(-\frac{\tau}{4t}\right), \tag{3.21}$$

whose Laplace transform is $\psi(s) = \exp\left(-\sqrt{\tau s}\right)$ (see Eq.(29.3.82) of Ref.[4]), which for $s \to 0$ follows $\psi(s) \cong 1 - \tau^{1/2}s^{1/2}$, as discussed above. The corresponding probabilities $\chi_n(t)$ are shown in Fig. 3.4. The Lévy–Smirnov PDF is an example of one-sided Lévy laws, discussed in Exercise 3.8 (see also Chapter 7), namely that with $\alpha = 1/2$.

A model in which a CTRW of such a type appears is a geometric comb model with infinite teeth (see Fig. 3.5), where we concentrate on the particle's displacement along the x-axis (backbone), and the waiting times correspond to the motion along the teeth. The waiting times in the teeth correspond to the return times of the walker to its initial position in the y-direction and are given by Eq.(2.15) (or its continuous-time analog). Thus, for a comb $\psi(t) \propto t^{-3/2}$.

Another example is: $\psi(t) = -\frac{d}{dt}\left(e^{t/\tau}\,\mathrm{erfc}\sqrt{t/\tau}\right)$, where $\mathrm{erfc}(x)$ is a complementary error function (see Eq.(7.1.2) of Ref.[4]). The asymptotic behavior of this $\psi(t)$ is given by Eq.(7.1.23) of Ref.[4], and is exactly the same as in our previous example, $\psi(t) \cong \frac{1}{2\sqrt{\pi}}\frac{\tau^{1/2}}{t^{3/2}}$. The corresponding survival probability $\Psi(t) = e^{t/\tau}\mathrm{erfc}\sqrt{t/\tau}$ has a

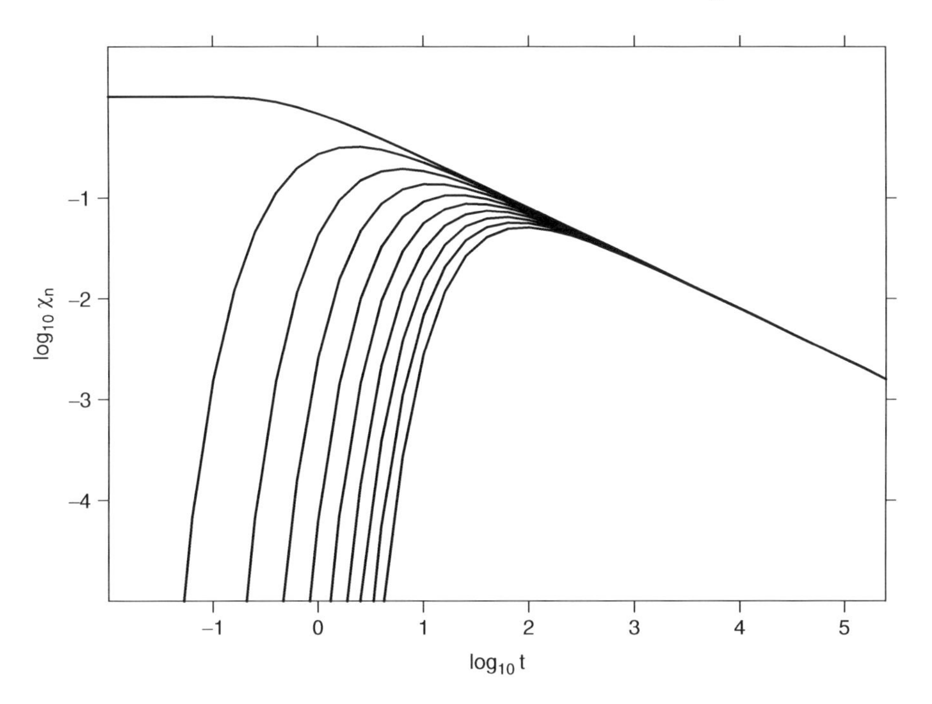

Fig. 3.4 *The first 10 functions $\chi_n(t)$ $(n = 0, 1 \ldots 9)$ for the waiting-time density Eq.(3.21) (here we take $\tau = 2$).*

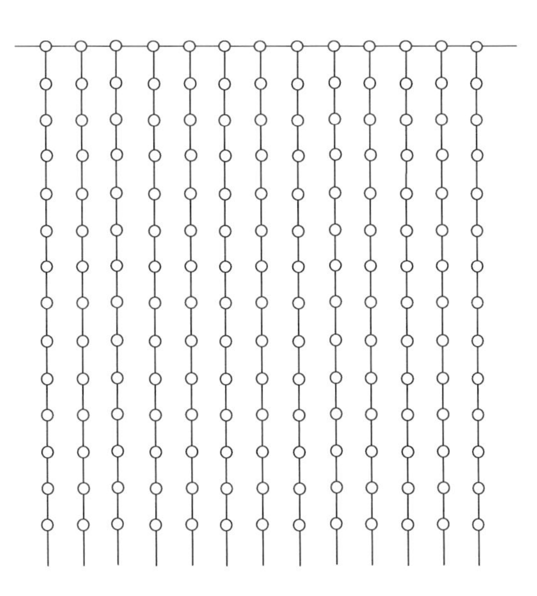

Fig. 3.5 *The displacement of a random walker in a horizontal direction on a comb structure can be considered as a CTRW with power-law waiting times due to trapping in the teeth.*

Laplace transform $\Psi(s) = \frac{1}{s+(s/\tau)^{1/2}}$ and is one of the representatives of the family of Mittag-Leffler functions $E_\alpha\left[-(t/\tau)^\alpha\right]$ (see Chapter 6), defined by their Laplace transform $f(s) = \frac{1}{s+\tau^{-\alpha}s^{1-\alpha}}$, again for $\alpha = 1/2$.

The one-sided Lévy distributions and the derivatives of Mittag-Leffler functions give us nice analytical models for waiting-time PDFs with heavy tails defined not only asymptotically but over the whole time domain. An additional advantage of them is the existence of effective algorithms generating corresponding random numbers (see Chapter 7).

Heavy-tailed waiting-time PDFs can also be considered as mixtures of exponential ones with different amplitudes and rates. Let us discuss the following example:

$$\psi(t) = \frac{1-b}{b}\sum_{j=1}^{\infty} b^j \beta^j e^{-\beta^j t},$$

which makes sense for $0 < b < 1$ and $\beta > 0$. This example resembles the Weierstrass function in Chapter 1. Let us consider $\psi(\beta t)$, which is given by

$$\psi(\beta t) = \frac{1-b}{b}\sum_{j=1}^{\infty} b^j \beta^j e^{-\beta^{j+1} t} = \frac{1}{b\beta}\frac{1-b}{b}\sum_{j=1}^{\infty} b^{j+1}\beta^{j+1} e^{-\beta^{j+1} t} = \frac{\psi(t)}{b\beta} - \frac{1-b}{b}e^{-\beta t}.$$

For t large enough the second term can be neglected, and we get

$$\psi(\beta t) = \frac{1}{b\beta}\psi(t).$$

The solution to this equation is a power law $\psi(t) \propto 1/t^{1+\alpha}$ with $\alpha = \frac{\ln b}{\ln \beta}$, which leads to distributions lacking the first moment for $\beta < b$.

The corresponding model can also be considered in the continuous case. Such a model arises when discussing subdiffusion in strongly disordered semiconductors [7].

Exercise 3.9 The waiting time t of a particle in a trap characterized by the mean waiting time τ follows the exponential distribution $\psi(t|\tau) = \frac{1}{\tau}\exp\left(-\frac{t}{\tau}\right)$. The characteristic sojourn time τ for a walker in a trap of energetic depth U is given by $\tau \propto \exp\left(\frac{U}{k_B T}\right)$, where k_B is the Boltzmann constant and T is the temperature. The distribution of the traps' depths follows the exponential law $p(U) \propto \exp\left(-\frac{U}{U_0}\right)$. Show that the PDF of characteristic waiting times τ in the traps follows a power law $p(\tau) \propto \frac{\tau_0}{\tau^{1+\alpha}}$ with $\alpha = \frac{k_B T}{U_0}$. Show that the waiting-time PDF $\psi(t)$ possesses the same power-law asymptotics.

The geometric analogy of the model of energetic disorder discussed above will be a comb model with teeth of random length. Note that the behavior of anomalous diffusion in the trap models depends on their spatial dimension (see Ref.[8] for a discussion).

3.5 Mean number of steps, MSD, and probability of being at the origin

Let us now consider the mean number of steps $\langle n(t)\rangle$ in a heavy-tailed case. Using Eqs.(3.17) and (3.20) we obtain $\langle n(s)\rangle \approx \frac{1}{\tau^\alpha s^{\alpha+1}}$ where only the terms of leading order are retained. Applying Tauberian theorems then gives

$$\langle n(t)\rangle = \frac{1}{\Gamma(1+\alpha)} \frac{t^\alpha}{\tau^\alpha}. \tag{3.22}$$

Contrary to the case of waiting-time PDFs possessing the first moment, whose mean number of steps always grows linearly in time, here we encounter a *sublinear* growth (recall that $0 < \alpha < 1$).

The mean number of steps in a heavy-tailed CTRW grows as $\langle n(t)\rangle \propto t^\alpha$. The corresponding step rate $k(t) = \frac{d}{dt}\langle n(t)\rangle$ decays with time.

Exercise 3.10 Calculate the mean number of steps done up to time t in a CTRW with the waiting-time PDF that asymptotically behaves as $\psi(t) \sim \frac{1}{t \ln^\beta t}$ $(\beta > 1)$ (see Exercise 3.4).

Exercise 3.11 The rate of taking a step is $k(t) = \sum_{n=0}^{\infty} \psi_n(t)$. Use Eq.(3.3) to obtain the closed expression for $k(t)$ in the Laplace domain. Show that for the exponential waiting-time PDF, Eq.(3.11), the rate $k(t)$ is constant and equals τ^{-1} for all $t > 0$. Show also that for a heavy-tailed distribution the rate decays with time as $k(t) \propto t^{\alpha-1}$. This finding exemplifies the property of *aging* of CTRW with heavy tails, discussed in Chapter 4.

Note the rate $k(t)$ at which steps are made is exactly $k(t) = \frac{d}{dt}\langle n(t)\rangle$.

Returning to Eq.(3.18) we immediately get the expression for the MSD under heavy-tailed waiting-time PDFs,

$$\langle x^2(t)\rangle = const \cdot \langle l^2\rangle \left(\frac{t}{\tau}\right)^\alpha,$$

which now increases more slowly than proportionally to the first power of time, which is a fingerprint of *subdiffusion*. The last expression can be rewritten in the form

$$\langle x^2(t)\rangle = \frac{1}{\Gamma(1+\alpha)} \frac{\langle l^2\rangle}{\tau^\alpha} t^\alpha = 2K_\alpha t^\alpha, \tag{3.23}$$

where $K_\alpha = \langle l^2\rangle / 2\tau^\alpha \Gamma(1+\alpha)$ is the generalized diffusion coefficient, having the dimension $[\mathrm{K}_\alpha] = \left[\frac{\mathrm{L}^2}{\mathrm{T}^\alpha}\right]$. In the case $\alpha \to 1$, Eq.(3.23) tends to that for MSD in normal diffusion, with the diffusion coefficient $K_1 = D = \frac{\langle l^2\rangle}{2\tau}$ (cf. Eq.(3.16)).

The MSD in a heavy-tailed CTRW grows as $\langle x^2(t) \rangle \propto t^\alpha$, and the probability of being at the origin decays as $P(0,t) \propto t^{-\alpha/2}$.

Let us now turn to the PDF of the particles' displacements in a CTRW with a heavy-tailed waiting-time distribution. Our model here will be analogous to that leading to Eq.(3.12) in the case of exponential waiting times. For a heavy-tailed waiting-time PDF we get

$$P(k,s) = \frac{\tau^\alpha s^{\alpha-1}}{\frac{k^2\sigma^2}{2} + \tau^\alpha s^\alpha}, \tag{3.24}$$

which reduces to Eq.(3.12) if we take $\alpha = 1$. Equation (3.24) can be of immediate use: It leads to the expression for the moments of the distribution and for the probability of finding a walker at the origin. The latter is given by the integral

$$P(0,s) = \frac{\tau^\alpha s^{\alpha-1}}{2\pi} \int_{-\infty}^{\infty} \frac{dk}{\frac{k^2\sigma^2}{2} + \tau^\alpha s^\alpha}.$$

Changing the variable of integration to $x = \sqrt{\sigma^2/2s^\alpha\tau^\alpha}k$ we get

$$P(0,s) = \frac{\tau^{\alpha/2}}{\sqrt{2}\pi\sigma s^{1-\alpha/2}} \int_{-\infty}^{\infty} \frac{dx}{x^2+1} = \frac{\tau^{\alpha/2}}{\sqrt{2}s^{1-\alpha/2}}$$

The inverse Laplace transform of this expression follows by use of Tauberian theorems, and gives asymptotically

$$P(0,t) = \frac{\tau^{\alpha/2}}{\sqrt{2}\sigma\Gamma(1-\alpha/2)} t^{-\alpha/2}.$$

Note that $P(0,t)$ is essentially of the order of $\langle x^2(t) \rangle^{-1/2}$ as given by Eq.(3.23). The height of the PDF of the walker's displacements in its center is of the order of its inverse width, as it has to follow from the normalization condition.

The probability of being at the origin can be obtained in yet another way, which is illustrated in Exercise 3.12 for a lattice random walk.

Exercise 3.12 According to Example 1.1 in Chapter 1, the probability of being at the origin after n steps behaves as $P_n(0) \sim 1/\sqrt{2\pi\sigma^2 n}$. Use Eq.(3.8) and the result for $\chi_n(s)$ given in Exercise 3.7 to rederive the expression for $P(0,t)$.

Hint: Consider $P(0,s)$ and approximate n as a continuous variable.

The inverse Fourier–Laplace transform of Eq.(3.24) leads to expressions involving so-called Fox functions, which in some cases can be reduced to simpler functions. Figure 3.6 presents the corresponding PDFs for $\alpha = 1/2$ for different times. Note in

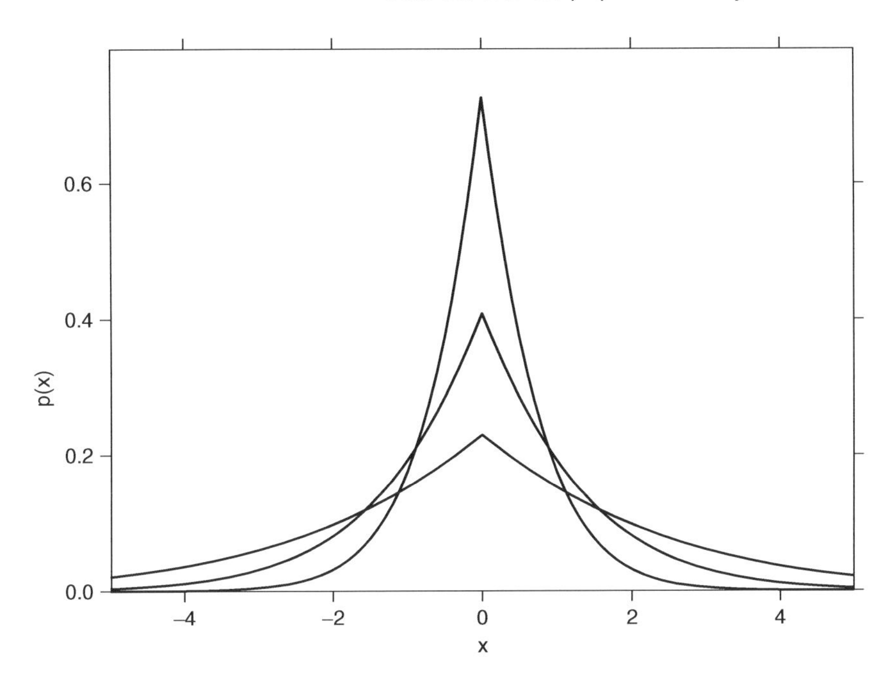

Fig. 3.6 *PDF of a CTRW with $\psi(t) \propto t^{-3/2}$ at three different times, $t_1 : t_2 : t_3 = 1 : 10 : 100$.*

particular the cusp at $x = 0$, which is typical for PDFs in CTRWs with $\alpha < 1$ in one dimension. The PDF $P(x,t)$ scales as a function of $x/t^{1/4}$, which essentially follows from its Fourier–Laplace representation.

Exercise 3.13 Show using Eq.(3.24) that the PDF $P(x,t)$ in a CTRW with a waiting-time distribution following a power law with exponent α scales as a function of $x/t^{\alpha/2}$, i.e., $P(x,t) = \frac{1}{t^{\alpha/2}} f\left(\frac{x}{t^{\alpha/2}}\right)$.

3.6 Other characteristic properties of heavy-tailed CTRW

In Chapter 2 we considered the return probabilities $F_n(\mathbf{0})$ for random walks on lattices, as well as the mean number of distinct sites visited, $\langle S_n \rangle$, as functions of the number of steps n. In Chapter 9 we show how important these properties are for the description of chemical reactions. Here it is appropriate to generalize the results of Chapter 2 for the case of CTRW. We make use of the idea of subordination connecting the internal clock of the walker (the number of steps n) with the physical time t. This connection is, however, different for the case of the first return time and for the mean number of steps. This difference corresponds to the difference between $\psi_n(t)$ and $\chi_n(t)$ discussed at the end of Sec. 3.1.

Let us start from a discussion of the mean number of sites visited *up to the time* t. The situation here is parallel to the calculation of $P(x,t)$, Eq.(3.8):

$$\langle S(t)\rangle = \sum_{n=0}^{\infty} \langle S_n\rangle \, \chi_n(t). \tag{3.25}$$

In three dimensions, when $\langle S_n\rangle \sim n$ and is given by Eq.(2.26) we thus have

$$\langle S(t)\rangle = \sum_{n=0}^{\infty} \frac{n}{P(\mathbf{0},1)} \chi_n(t) = \frac{\langle n(t)\rangle}{P(\mathbf{0},1)},$$

with $\langle n(t)\rangle$ being the mean number of steps performed up to the time t defined in Exercise 3.3. For the waiting-time distributions possessing the mean waiting time τ, this is simply $\langle n(t)\rangle = t/\tau$; for the heavy-tailed power laws it is $\langle n(t)\rangle = \frac{1}{\Gamma(1+\alpha)}\frac{t^\alpha}{\tau^\alpha}$, Eq.(3.22).

In the one-dimensional case $\langle S(t)\rangle$ is defined by the weighted sum of $\langle S_n\rangle \cong \sqrt{(2/\pi)n}$. Here we can use the asymptotic form $\chi_n(t) \approx \frac{t}{\alpha\tau} n^{-\frac{1}{\alpha}-1} L_\alpha\left(\frac{t}{\tau n^{1/\alpha}}\right)$ as discussed in Exercise 3.8 and approximate the sum by an integral

$$\langle S(t)\rangle \approx \sqrt{\frac{2}{\pi}} \int_0^\infty \frac{t}{\alpha\tau} n^{-\frac{1}{\alpha}-\frac{1}{2}} L_\alpha\left(\frac{t}{\tau n^{1/\alpha}}\right) dn = \sqrt{\frac{2}{\pi}} \frac{1}{\Gamma(1+\alpha)} \left(\frac{t}{\tau}\right)^{\frac{\alpha}{2}} \tag{3.26}$$

(the result is easily obtained by passing to a Laplace representation in t as is done in Exercise 3.8, performing the integration in n, and then returning to the time domain). Alternatively, we can use the formula

$$\int_0^\infty y^\eta L_\alpha(y) dy = \frac{\Gamma(1-\eta/\alpha)}{\Gamma(1-\eta)},$$

$(-\infty < \eta < \alpha)$ (see Ref.[9]).

Note that the exponent governing the time dependence of $\langle S(t)\rangle$ is a product of two exponents, one governing the behavior of $\langle S_n\rangle \propto n^\gamma$ as a function of n, and depending on the dimensionality of the walk ($\gamma = 1$ for $d = 3$ and $\gamma = 1/2$ for $d = 1$), and another one governing the behavior of $\langle n(t)\rangle$ as a function of t ($\langle n(t)\rangle \propto t^\alpha$ in both cases, provided $0 < \alpha < 1$).

The same overall behavior is observed for $P(\mathbf{0},t)$, generalizing the result of Exercise 3.9: It is given by the weighted sum of $P_n(\mathbf{0}) \propto \frac{1}{\left(\sqrt{2\pi\sigma^2 n}\right)^d}$, and thus leads to

$$P(\mathbf{0},t) \propto t^{-d\alpha/2}. \tag{3.27}$$

In several models considered in Chapter 10, for example for random walks on fractal substrates, we will encounter random-walk processes in which the MSD $\langle x^2(n)\rangle$ grows as $\langle x^2(n)\rangle \propto n^\gamma$ (with $\gamma \neq 1$) as a function of the number of steps. The corresponding CTRWs then show a similar subordinated behavior, namely

$$\langle x^2(t) \rangle \propto t^{\alpha\gamma}. \tag{3.28}$$

Exercise 3.14 Calculate the mean number of distinct sites visited over a long time t by a random walker on a two-dimensional lattice, where $\langle S_n \rangle \cong \pi n / \ln n$. Consider the cases of the exponential waiting-time PDF and of the power-law waiting-time PDF.

Hint: Using Tauberian theorems is helpful. Do not forget the slowly varying function!

The situation with the first return time is different, since the first return occurs exactly on the time of a step, excluding the possibility of arriving earlier and waiting. The PDF of the time of the first return is given by

$$F(t,0) = \sum_{n=0}^{\infty} F_n(0) \psi_n(t).$$

In one dimension $F_n(0)$ is given by Eq.(2.15), $F_n(0) = \frac{1}{\sqrt{2\pi}} n^{-3/2}$. Noting that $\psi_n(s) = \chi_n(s) \frac{s}{1-\psi(s)}$ we find that for small s this quantity behaves as $\psi_n(s) \approx \exp(-n\tau^\alpha s^\alpha)$. Repeating the tricks of Exercise 3.14, we get $F(t,0) \propto t^{-1-\alpha/2}$, which is *not* a product of the exponents of the n- and t-dependences.

References

[1] E. Montroll and G.H. Weiss. *J. Math. Phys.* **6**, 167–181 (1965)
[2] G. Arfken. *Mathematical Methods for Physicists*, Academic Press: Boston, 1985
[3] G. Doetsch. *Introduction to the Theory and Application of the Laplace Transformation*, Berlin: Springer, 1974
[4] M. Abramovitz and I.A. Stegun. *Handbook of Mathematical Functions*, New York: Dover, 1972
[5] W. Feller. *An Introduction to Probability Theory and Its Applications*, New York: Wiley, 1971 (Tauberian theorems are discussed in Vol. 2 Chap. XIII, Sec. 5.)
[6] A. Blumen, J. Klafter, and G. Zumofen. "Reactions in Disordered Media Modelled by Fractals," in: Pietronero, L., and Tosatti, E., eds., *Fractals in Physics*, Amsterdam: North-Holland, 1986, pp. 399–408
[7] H. Scher, M.F. Shlesinger, and J.T. Bendler. *Physics Today* **44** (1), 26 (1991)
[8] J.P. Bouchaud and A. Georges. *Phys. Repts.* **195**, 127–293 (1990)
[9] K.-I. Sato. *Lévy Processes and infinitely Divisible Distributions*, Cambridge: Cambridge University Press, 2002

4
CTRW and aging phenomena

"It is as if an ox had passed through a window screen: Its head, horns, and four hooves have all passed through; why can't the tail pass through?"

Wuzu Fayan

As we have seen in Chapter 3 the *rate of jumps* in a CTRW with power-law waiting-time PDFs decays with time. This property of the heavy-tailed waiting-time PDFs has far-reaching consequences for the kinetics of such processes. In the present chapter we discuss some of them, namely ergodicity breaking and the decay of linear response.

4.1 When the process ages

Up to now we have considered situations when a physical clock was started at the first step of the process. In physical terms this means that we start the measurement on the system immediately after its preparation in the present state. Let us now concentrate on the MSD of a random walker during the time interval Δt starting not at $t = 0$ but at $t = t_a > 0$, well after the system has been prepared. Therefore the measured MSD is

$$\left\langle x^2(\Delta t) \right\rangle = \left\langle (x(t_a + \Delta t) - x(t_a))^2 \right\rangle . \tag{4.1}$$

We start with a simple example of a one-dimensional decoupled, symmetric CTRW with a heavy-tailed waiting-time PDF and assume that the random walker is initiated at time $t = 0$; however, the measurement (observation time) starts not at time $t = 0$ but at some later instant of time $t = t_a$. The time t_a is referred to as the aging time. Let the walker's position at $t = t_a$ be $x(t_a)$. From this time on we follow the walker and are interested in its MSD given by Eq.(4.1). According to Eq.(1.11) (and parallel to Eq.(3.18)) this MSD is $\langle x^2(\Delta t)\rangle = \langle l^2\rangle\langle n(\Delta t, t_a)\rangle$, where $n(\Delta t, t_a)$ is the number of steps done between t_a and $t_a + \Delta t$. Note that $\langle n(\Delta t, t_a)\rangle = \langle n(t_a + \Delta t) - n(t_a)\rangle = \langle n(t_a + \Delta t)\rangle - \langle n(t_a)\rangle$, so that

$$\left\langle x^2(\Delta t) \right\rangle = \left\langle l^2 \right\rangle \left[\langle n(t_a + \Delta t)\rangle - \langle n(t_a)\rangle\right] . \tag{4.2}$$

The analysis of this expression reveals the difference between two situations:

(i) Waiting-time PDF for which the mean waiting time $\langle t\rangle = \tau$ exists and
(ii) heavy-tailed power-law waiting-time PDFs.

Taking each of these in turn:

(i) If the mean waiting time τ exists (say, in the case of the exponential waiting-time PDF), we have $\langle n(t)\rangle \cong t/\tau$ (see Chapter 3), so that $\langle x^2(\Delta t)\rangle \propto \langle l^2\rangle \Delta t/\tau$ and is independent of t_a. This situation corresponds to the time-homogeneous behavior.

(ii) In the case of a power-law waiting-time PDF, Eq.(3.19), we have $\langle n(t)\rangle \propto t^\alpha$ (see Eq.(3.21)), so that $\langle x^2(\Delta t)\rangle \propto \langle l^2\rangle \left[(t_a+\Delta t)^\alpha - t_a^\alpha\right]$. We see that the MSD during the time Δt depends explicitly on the time t_a when the observation started, i.e., on the age of the process at the beginning of measurement. Looking at the asymptotic behavior of the MSD we obtain the following limits:

- For $\Delta t \gg t_a$, we get $\langle x^2(\Delta t)\rangle \propto \Delta t^\alpha$, i.e., if the observation time is much longer than the aging time, the latter plays no role and can be neglected. We essentially recover the results of Chapter 3.
- For $\Delta t \ll t_a$, we can expand the first expression in a Taylor series and get $\langle x^2(\Delta t)\rangle \propto t_a^{\alpha-1}\Delta t$. This is a very interesting result: If the observation time is much shorter than the aging time, we observe normal diffusion in a system showing otherwise anomalous behavior. The diffusion coefficient is then proportional to $t_a^{\alpha-1}$, i.e., decays when the aging time increases (since $\alpha < 1$).

This simple discussion serves as an introduction to aging phenomena. Two situations will be considered: the one already discussed, namely, the initial aging condition, and another one corresponding to the behavior of the system's linear response to an external field.

Exercise 4.1 Obtain the MSD $\langle x^2(\Delta t)\rangle$ after the aging time t_a for a CTRW with the waiting-time PDF that asymptotically behaves as $\psi(t) \sim \frac{1}{t\ln^\beta t}$ $(\beta > 1)$ (see Exercise 3.10).

4.2 Forward waiting time

In order to address the dependence on the aging time, we have to know how long the walker has to wait from its initiation time until making its first step after the observation started. The distribution of this so-called forward waiting time may differ from the distribution of waiting times for all subsequent steps. The PDF of the forward waiting time will be denoted by $\psi_1(t, t_a)$. Let us calculate this PDF provided that the waiting-time PDF $\psi(t)$ and the time t_a are known. In Chapter 3, the aging time was assumed to be zero, and for this case $\psi_1(t, t_a) = \psi(t)$. We first give a general derivation and then discuss how we can understand the result in a simple way.

Let n be the number of steps performed by the walker until the observation started at t_a, so that t_a falls in the interval between the time $T_n = \sum_{i=1}^{n} t_n$ (when the n-th step was made) and T_{n+1}. Given T_n, we can easily calculate the PDF of the forward waiting time t by noting that the actual waiting time until the next step is $t_a - T_n + t$. Thus, the conditional probability density of the forward waiting time t provided that exactly n steps took place before t_a reads

$$\phi_n(t) = \int_0^{t_a} \psi_n(t')\psi(t_a - t' + t)dt',$$

where $\psi_n(t)$ is defined in Chapter 3 and is the PDF of T_n. The value of $\psi_1(t, t_a)$ follows by summing over n:

$$\psi_1(t, t_a) = \sum_{n=0}^{\infty} \phi_n(t) = \int_0^{t_a} \left(\sum_{n=0}^{\infty} \psi_n(t') \right) \psi(t_a - t' + t)dt'.$$

The sum $\sum_{n=0}^{\infty} \psi_n(t') = k(t')$ is the rate of jumps defined in Exercise 3.11, and the expression for the forward waiting-time PDF is therefore

$$\psi_1(t, t_a) = \int_0^{t_a} k(t')\, \psi(t_a - t' + t)dt'. \tag{4.3}$$

The forward waiting-time PDF is thus a convolution of the rate $k(t)$ and the waiting-time PDF $\psi(y)$ depending on the "shifted" time variable $y = t + t_a$. Equation (4.3) can hardly be simplified, but can still be evaluated for some cases.

Exercise 4.2 Evaluate $\psi_1(t, t_a)$ explicitly for the case with exponential waiting-time PDF $\psi(t) = \tau^{-1} \exp(-t/\tau)$. Show that for this special case $\psi_1(t, t_a)$ coincides with $\psi(t)$.

4.2.1 The inspection paradox

As the first example let us consider the situation of processes possessing mean waiting times. The Laplace transform of $k(t)$ reads (see Exercise 3.11):

$$k(s) = \frac{1}{1 - \psi(s)}.$$

For waiting-time PDFs possessing a mean waiting time τ we have $\psi(s) = 1 - s\tau + \ldots$ so that the Laplace transform of $k(t)$ tends to $1/s\tau$ for small s. This means that $k(t)$ itself tends for large t to a finite limit $k(t) \to \tau^{-1}$. Then, for t_a large enough we get

$$\psi_1(t, t_a) = \frac{1}{\tau} \int_0^{t_a} \psi(t_a - y + t)dy = \frac{1}{\tau} \int_t^{t_a+t} \psi(z)dz = \frac{1}{\tau} [F(t + t_a) - F(t)],$$

where $F(t) = \int_0^t \psi(t')dt'$ is the cumulative distribution function of waiting times. For very long aging times leading to equilibration, we get $F(t + t_a) \to 1$ and

$$\psi_1^{\text{eq}}(t) = \tau^{-1}[1 - F(t)] = \tau^{-1}\Psi(t). \tag{4.4}$$

The mean waiting time for the first step (mean forward waiting time) can be obtained by taking the Laplace transform of this expression:

$$\psi_1^{\text{eq}}(s) = \tau^{-1}\Psi(s) = \frac{1 - \psi(s)}{s\tau}.$$

Let us now assume that the waiting-time density $\psi(t)$ possesses two moments, i.e., its Laplace transform is $\psi(s) = 1 - s\tau + s^2\langle t^2\rangle/2 - \ldots$. In this case the mean forward waiting time (the mean time of waiting from the beginning of observation until a new step) $\langle t_f\rangle = \int_0^\infty t\psi_1^{eq}(t)dt$ follows from the expansion of $\psi_1^{\text{eq}}(s) = 1 - \left[\langle t^2\rangle/2\tau\right]\ s + \ldots$ and reads

$$\langle t_f\rangle = \langle t^2\rangle/2\tau.$$

Let us now consider examples shedding light on this behavior for three special cases.

If the steps follow periodically with waiting time τ between them, then $\psi(t) = \delta(t - \tau)$ so that the mean forward waiting time $\langle t_f\rangle = \tau/2$. This is exactly what our intuition suggests.

The case of the exponential waiting-time PDF has already been considered in Exercise 4.2. For this case the PDF of the forward waiting time is the same as for all other waiting times, and the mean forward waiting time is the same as the mean waiting time τ.

Let us now consider the waiting-time PDF $\psi(t) = (t/\tau^2)\exp(-t/\tau)$ whose mean waiting time is again τ. When now calculating the mean forward waiting time, we obtain $(3/2)\tau$, which is *larger* than τ. This result may seem paradoxical, since the observation of the process starts *between* the two events, so that we expect that the mean forward waiting time will be *less* than τ. This is one of the well-known paradoxes of probability theory, called the inspection paradox or the waiting-time paradox.

After the general result, Eq.(4.4), is obtained and illustrated by examples, a simple explanation is in place. Imagine that our process has started long ago ($t_a >> \tau$) so that we have to deal with an equilibrated process, namely, any correlation between the beginning of the process and the inspection (beginning of observations) is lost. Thus we can assume that the inspection takes place at random, absolutely independent of steps. Geometrically (see Fig. 4.1) this corresponds to throwing a point representing the inspection time onto the t-axis and measuring the time until the next step takes place. The probability that the random point lands within the interval of duration t_l is proportional to the length t_l of the interval (longer intervals are sampled more probably than short ones; here lies the explanation of the paradox) and to the probability of finding an interval of duration from t_l to $t_l + dt_l$ among all intervals, so that $p(t_l)dt_l \propto t_l\psi(t_l)dt_l$. To obtain a proper normalization, we take $p(t_l) = t_l\psi(t_l)/\int_0^\infty t'\psi(t')dt' = (t_l/\tau)\,\psi(t_l)$. Now, within the interval of length t_l the inspection time can again be considered to be placed at random and thus equally distributed with the probability

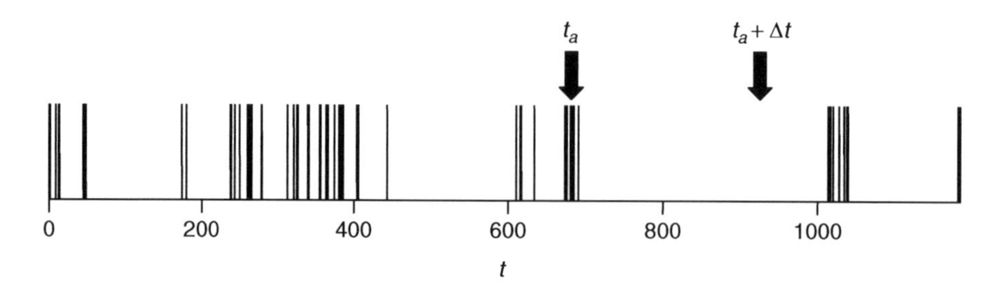

Fig. 4.1 *A realization of jump times in a CTRW with power-law distribution of waiting times (the same as in Fig. 3.3), shedding light on the aging initial condition.*

density $w = 1/t_l$. The PDF of the forward waiting time t within the interval of length t_l is therefore $1/t_l$ provided that $t < t_l$. This has to be integrated over the distribution of the interval lengths t_l, which leads to $\psi_1^{\text{eq}}(t) = \int_t^\infty \frac{1}{t_l} \frac{t_l}{\tau} \psi(t_l) dt_l = \tau^{-1} [1 - F(t)]$, i.e., exactly Eq.(4.4).

Now we can discuss the nature of the inspection paradox considered above. Although the waiting time until the next step is always shorter than the time interval between the two corresponding steps (when the inspection time does not coincide with the beginning of the interval between steps), this may be more than compensated for by the fact that the inspection time falls with larger probability in the longer intervals between steps, which are thus overrepresented.

For the waiting-time distributions possessing the first but no second moment, the result Eq.(4.4) leads to infinite mean forward waiting times: Such distributions show the extreme form of the inspection paradox. For the waiting-time distributions lacking the first moment the whole approach based on Eq.(4.4) is no longer applicable, and we have to use the general scheme based on Eq.(4.3). We now proceed by obtaining the asymptotic results for the case of a power-law waiting-time PDF also lacking the first moment.

4.2.2 Forward waiting-time PDF in Laplace domain

The Laplace transform of the shifted waiting-time PDF $\psi(y + t)$ in y reads:

$$\int_0^\infty \psi(y+t) e^{-uy} dy = e^{ut} \left[\psi(u) - \int_0^t e^{-uy} \psi(y) dy \right],$$

so that the Laplace transform of $\psi_1(t, t_a)$ in its second temporal variable t_a is given by

$$\psi_1(t, u) = \frac{e^{ut} \left[\psi(u) - \int_0^t e^{-uy} \psi(y) dy \right]}{1 - \psi(u)}. \tag{4.5}$$

Now it is very easy to obtain the Laplace transform of the forward waiting-time PDF also in its first temporal variable. We leave it to the reader.

Exercise 4.3 Evaluate $\psi_1(s,u) = \int_0^\infty \int_0^\infty \psi_1(t,t_a) e^{-st} e^{-ut_a}\, dt\, dt_a$. Show that

$$\psi_1(s,u) = \frac{1}{1-\psi(u)} \frac{\psi(u)-\psi(s)}{s-u}. \tag{4.6}$$

Hint: First rewrite Eq.(4.5) as $\psi_1(t,u) = \frac{1}{1-\psi(u)} e^{ut} \int_t^\infty e^{-uy}\psi(y)\, dy$ and use partial integration when evaluating $\psi_1(s,u)$.

Exercise 4.4 Prove the normalization of $\psi_1(t,t_a)$ using Eq.(4.6).

Hint: First take $s \to 0$ and then perform the inverse Laplace transform in u.

Exercise 4.5 There is a simpler way to obtain Eq.(4.6) immediately from Eq.(4.3). To do this show that the Laplace transform of any function $f(t_1+t_2)$ in both temporal variables t_1 and t_2 reads $f(s,u) = \frac{f(u)-f(s)}{s-u}$.

4.2.3 Power-law waiting-time distributions

Let us now consider the case of power-law waiting-time distribution $\psi(t) \propto \tau^\alpha t^{-1-\alpha}$ (Eq.(3.19)) lacking the first moment. The Laplace transform of such a distribution for small s is given by $\psi(s) \cong 1 - \tau^\alpha s^\alpha$ (Eq.(3.20)).

We now evaluate first $\psi_1(t,s)$ using Eq.(4.5) and then perform the inverse Laplace transform to obtain $\psi_1(t,t_a)$. We first rewrite Eq.(4.5) as

$$\psi_1(t,s) = \frac{1}{1-\psi(s)} e^{st} \int_t^\infty e^{-sy}\psi(y) dy \cong (s\tau)^{-\alpha} e^{st} \int_t^\infty e^{-sy} \frac{\alpha\tau^\alpha}{\Gamma(1-\alpha) y^{1+\alpha}} dy$$

and get for large t

$$\psi_1(t,s) = \frac{\alpha e^{st}}{\Gamma(1-\alpha)} \int_{st}^\infty e^{-z} z^{-1-\alpha} dz = \frac{\alpha e^{ts}}{\Gamma(1-\alpha)} \Gamma(-\alpha, st), \tag{4.7}$$

where $\Gamma(\beta,\, x)$ is the incomplete Γ-function (Eq.(6.4.3) of Ref. [1]). Now we can easily take an inverse Laplace transform of Eq.(4.7) with respect to s (see Eq.(2.10.16) of Ref. [2]):

$$\psi_1(t,t_a) = \frac{\alpha}{\Gamma(1-\alpha)\Gamma(1+\alpha)} \left(\frac{t_a}{t}\right)^\alpha \frac{1}{t+t_a} = \frac{\sin \pi\alpha}{\pi} \left(\frac{t_a}{t}\right)^\alpha \frac{1}{t+t_a}. \tag{4.8}$$

To obtain the latter expression, we have used Eqs.(6.1.15) and (6.1.17) of Ref. [1]. Note that for large t_a the forward waiting-time PDF $\psi_1(t,t_a)$ shows two distinct types of behavior: For $t < t_a$ it decays as $t^{-\alpha}$, which is quite a slow decay; while for $t > t_a$ the decay follows $\psi_1(t, t_a) \propto t^{-1-\alpha}$, which is a pattern of the decay of $\psi(t)$.

4.3 PDF of the walker's positions

To take it further and obtain the PDF of the particles' displacements in the decoupled aging CTRW we return to Eq.(3.8) and note that

$$P(x, \Delta t; t_a) = \sum_{n=0}^{\infty} P_n(x)\chi_n(\Delta t; t_a), \tag{4.9}$$

where we take into account the fact that the number of steps performed during the time interval Δt depends on the aging time t_a. Here $\chi_n(\Delta t; t_a)$ is the probability of making exactly n steps during the time interval from t_a up to $t_a + \Delta t$ in a system that has aged up to time t_a. In the Fourier–Laplace domain, $P(k, s; t_a) = \int_{-\infty}^{\infty} dx \int_0^{\infty} d(\Delta t)\, e^{ikx} e^{-s\Delta t} P(x, \Delta t; t_a)$, Eq.(4.9) reads

$$P(k, s; t_a) = \sum_{n=0}^{\infty} \lambda^n(k)\chi_n(s; t_a). \tag{4.10}$$

Our next step will be to calculate $\chi_n(s; t_a)$. Thus $\chi_0(\Delta t; t_a)$, giving the probability of making no steps during the interval from t_a up to $t_a + \Delta t$, is given by

$$\chi_0(\Delta t, t_a) = 1 - \int_0^{\Delta t} \psi_1(t, t_a)dt,$$

parallel to the expression for $\chi_0(t) = \Psi(t)$ in Chapter 3. Moreover,

$$\chi_1(\Delta t, t_a) = \int_0^{\Delta t} \psi_1(t, t_a)\Psi(\Delta t - t)dt$$

and all the subsequent $\chi_n(\Delta t; t_a)$ are given by the double convolution

$$\chi_n(\Delta t, t_a) = \int_0^{\Delta t} \left[\int_0^{t} \psi_1(t_1, t_a)\psi_{n-1}(t - t_1)dt_1 \right] \Psi(\Delta t - t)dt.$$

This equation is an analog of Eq.(3.6), with the only difference being that, owing to a special role of the first step, we replace $\psi_n(t)$ by a convolution $\psi_1(t_1, t_a)$ and $\psi_{n-1}(t)$. In the Laplace domain the expressions for $\chi_n(s; t_a)$ read:

$$\chi_0(s; t_a) = \frac{1 - \psi_1(s; t_a)}{s}$$

and

$$\chi_n(s;t_a) = \psi_1(s;t_a)\psi^{n-1}(s)\frac{1-\psi(s)}{s} \tag{4.11}$$

for $n > 0$.

Exercise 4.6 The mean number of steps done during the time interval Δt is given by $\langle n(\Delta t;t_a)\rangle = \sum_{n=1}^{\infty} n\chi_n(\Delta t;t_a)$. Show that $\langle n(\Delta t,t_a)\rangle = \langle n(t_a+\Delta t)\rangle - \langle n(t_a)\rangle$, where $\langle n(t)\rangle$ is the mean number of steps done up to time t in a non-aged (normal) random walk.

Inserting the expressions for $\chi_n(s;t_a)$ into Eq.(4.10) we get

$$\begin{aligned} P(k,s;t_a) &= \frac{1-\psi_1(s;t_a)}{s} + \psi_1(s;t_a)\frac{1-\psi(s)}{s}\sum_{n=1}^{\infty}\lambda^n(k)\psi^{n-1}(s) \\ &= \frac{1-\psi_1(s;t_a)}{s} + \frac{1-\psi(s)}{s}\frac{\psi_1(s;t_a)}{\psi(s)}\sum_{n=1}^{\infty}\lambda^n(k)\psi^n(s) \\ &= \frac{1-\psi_1(s;t_a)}{s} + \frac{1-\psi(s)}{s}\frac{\psi_1(s;t_a)}{\psi(s)}\frac{\lambda(k)\psi(s)}{1-\lambda(k)\psi(s)} \\ &= \frac{1-\psi_1(s;t_a)}{s} + \frac{1-\psi(s)}{s}\frac{\lambda(k)\psi_1(s;t_a)}{1-\lambda(k)\psi(s)}. \end{aligned} \tag{4.12}$$

This equation reduces to Eq.(3.10) if $\psi_1(s;t_a)$ is equal to $\psi(t)$. Note that this $P(k,s;t_a)$ is connected to that for non-aged (normal) CTRW $P(k,s)$ (Eq.(3.10)), via

$$P(k,s;t_a) = \frac{1-\psi_1(s;t_a)}{s} + \psi_1(s;t_a)\lambda(k)P(k,s). \tag{4.13}$$

Note that $\psi_1(s;t_a)\lambda(k)$ is the Fourier–Laplace transform of the joint PDF of the displacement and waiting time in the first step, which is denoted $p_1(x,t;t_a)$. Therefore Eq.(4.12) corresponds in the space–time domain to the following structure:

$$P(x,\Delta t;t_a) = \chi_0(\Delta t;t_a)\delta(x) + \int_{-\infty}^{\infty} dy \int_0^{\Delta t} dt\ p_1(y,t;t_a)P(x-y,\Delta t - t).$$

Let us now turn to a discussion of the second moment of aging CTRW, which was the starting point of the present chapter.

Exercise 4.7 Consider the case of $\psi(t)$ possessing the first moment τ. In the equilibrated situation the forward waiting-time PDF is given by Eq.(4.4). Show that in this case the $P(k,\ s;t_a \to \infty)$ for small s and k is given exactly by Eq.(3.10).

Let us consider the behavior of the MSD, which can be obtained either by taking the second derivative in k in Eq.(4.12) or by using Eq.(4.2) and the expression for $\langle n(t)\rangle$. We turn to the Laplace representation and recall that for the case when $\psi_1(t,\ t_a) = \psi(t)$, $\langle n(t)\rangle$ is given by Eq.(3.17), so that

$$\langle x^2(s)\rangle = \langle l^2\rangle \frac{\psi(s)}{s\,[1-\psi(s)]}.$$

In the case when $\psi_1(t,\ t_a) \neq \psi(t)$ we get

$$\langle x^2(s)\rangle = \langle l^2\rangle \frac{\psi_1(s,t_a)}{s\,[1-\psi(s)]}, \tag{4.14}$$

which differs from the previous equation by replacing $\psi(s)$ by $\psi_1(s,\ t_a)$ in the enumerator.

Exercise 4.8 Derive Eq.(4.14) using Eq.(4.11) and the trick from Exercise 3.3.

Equation (4.14) is especially interesting for the case of equilibrated CTRW, when $\psi_1(s,\ t_a \to \infty)$ follows from Eq.(4.4): $\psi_1(s,\ t_a \to \infty) = \psi_1^{\mathrm{eq}}(s) = \frac{1-\psi(s)}{s\tau}$ so that $\langle x^2(s)\rangle = \langle l^2\rangle/s^2\tau$ and $\langle x^2(\Delta t)\rangle = \langle l^2\rangle\, t/\tau$. In this case the MSD follows normal diffusion not only asymptotically but from the very beginning of observation.

4.4 Moving time averages

Up to now, the means, e.g., $\langle \Delta x^2(t)\rangle = \langle \Delta x^2(t)\rangle_{ens} = \langle (x(t)-x(0))^2\rangle_{ens}$, have been considered as ensemble averages, i.e., as mean values over many independent realizations of the corresponding process, which is implied by our probabilistic definition

$$\left\langle \Delta x^2(t)\right\rangle_{ens} = \int_{-\infty}^{\infty} (x(t)-x(0))^2\, P(x,t)dx.$$

In many experiments, especially those using single-particle tracking, moving time averages are applied. Having one long trajectory (realization) of the process $x(t)$ of the overall duration T, we define

$$\left\langle \Delta x^2(t)\right\rangle_T = \frac{1}{T-t}\int_0^{T-t} (x(t+\tau)-x(\tau))^2\, d\tau.$$

For normal diffusion, $\langle \Delta x^2(t)\rangle_{ens} = \langle \Delta x^2(t)\rangle_T$ provided that T is large enough, which can be referred to as an ergodic property. For CTRW with waiting-time PDFs lacking the mean waiting time, such as those of Eq.(3.19) or of Exercise 3.10, this is no longer the case. Results of numerical simulations of CTRW following the waiting-time PDF (Eq.(3.19)), are shown in Fig. 4.2. The algorithms used to simulate such waiting-time distributions are discussed in Chapter 7.

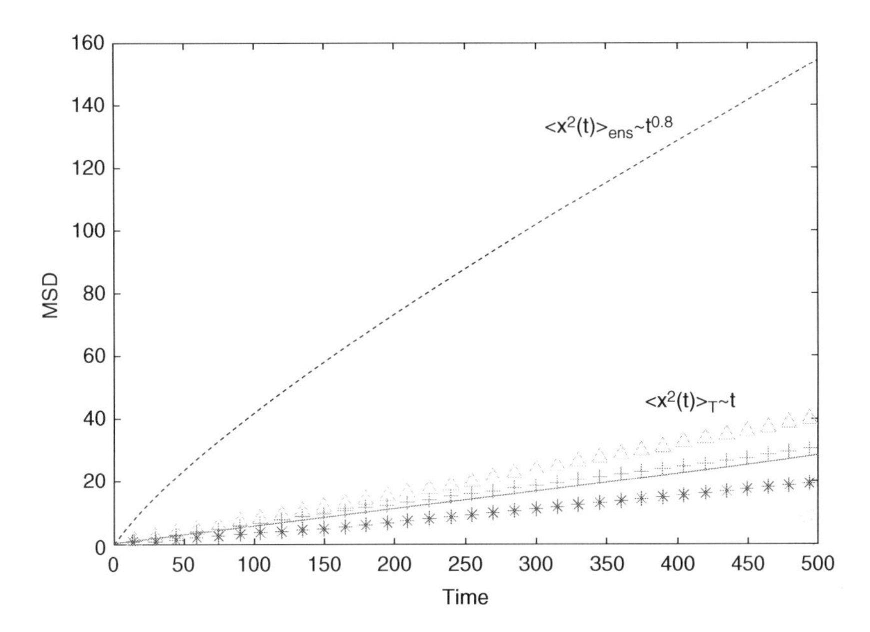

Fig. 4.2 *The ensemble-averaged MSD in the CTRW with $\psi(t) \propto t^{-1.8}$ (upper curve) and three different realizations of moving time averages (where the straight line indicates the ensemble average of moving time averages [3]).*

The conclusions we draw from this figure are as follows: While the ensemble-averaged MSD, $\langle \Delta x^2(t) \rangle_{ens}$, shows the power-law behavior $\langle \Delta x^2(t) \rangle \propto t^\alpha$, as discussed in Chapter 3, the time-averaged curves $\langle \Delta x^2(t) \rangle_T$ for different trajectories show very different behavior. In each realization the behavior of $\langle \Delta x^2(t) \rangle_T$ as a function of t is approximately linear, as in the case for *normal* diffusion. However, the diffusion coefficients in different realizations vary strongly. The difference between the ensemble and the time averages demonstrates a *non-ergodic* property.

To understand this non-ergodicity, it is instructive to consider an additional ensemble average over many trajectories of the process. Since the two averaging procedures $\langle \ldots \rangle_{ens}$ and $\langle \ldots \rangle_T$ involve independent integrations over different variables, x and t, they are interchangeable, so that

$$\left\langle \left\langle \Delta x^2(t) \right\rangle_T \right\rangle_{ens} = \left\langle \left\langle \Delta x^2(t) \right\rangle_{ens} \right\rangle_T = \frac{1}{T-t} \int_0^{T-t} \left\langle (x(t+\tau) - x(\tau))^2 \right\rangle_{ens} d\tau.$$

Now we use the result of Eq.(4.2) showing that $\langle (x(t+\tau) - x(\tau))^2 \rangle_{ens} = \langle l^2 \rangle (\langle n(t+\tau) \rangle - \langle n(t) \rangle)$ and that of Eq.(3.22) giving us $\langle n(t) \rangle = \frac{1}{\Gamma(1+\alpha)} \frac{t^\alpha}{\tau^\alpha}$. We therefore obtain

$$\left\langle \left\langle \Delta x^2(t) \right\rangle_T \right\rangle_{ens} = \frac{1}{\Gamma(1+\alpha)} \frac{\langle l^2 \rangle}{\tau^\alpha} \frac{1}{T-t} \int_0^{T-t} \left[(t+\tau)^\alpha - \tau^\alpha \right] d\tau,$$

which for $t << T$ can be approximated as

$$\left\langle \langle \Delta x^2(t) \rangle_T \right\rangle_{ens} = 2K_\alpha \frac{t}{T^{1-\alpha}}.$$

The combination $\frac{1}{2\Gamma(1+\alpha)} \frac{\langle l^2 \rangle}{\tau^\alpha}$ is identified with the generalized diffusion coefficient K_α of Eq.(3.23). We see that such a double average behaves as if diffusion were normal, $\left\langle \langle \Delta x^2(t) \rangle_T \right\rangle_{ens} = 2D_{eff} t$, and that the effective diffusion coefficient depends explicitly on the averaging time (trajectory length) T. The subdiffusive nature (for $\alpha < 1$) evident in the ensemble average appears here through the dependence of D_{eff} on the length of the averaging interval: D_{eff} decreases with increasing T. The dependence on T vanishes only for $\alpha = 1$ (normal diffusion) when the ergodicity is recovered.

Exercise 4.9 We return again to the ultra-slow CTRW of Exercise 3.10, that with $\psi(t) \sim \frac{1}{t \ln^\beta t} (\beta > 1)$. Show that in this case the double average $\left\langle \langle \Delta x^2(t) \rangle_T \right\rangle_{ens}$ behaves as $\left\langle \langle \Delta x^2(t) \rangle_T \right\rangle_{ens} \sim \frac{\ln^{\beta-1}(T)}{T} t$.

As discussed in Chapter 3, in several cases, for example for random walks on fractal substrates, the MSD $\langle x^2(n) \rangle$ grows as a function of the number of steps not linearly but according to $\langle x^2(n) \rangle \propto n^\gamma$ (with $\gamma \neq 1$). When now performing CTRW on such structures the non-ergodicity mentioned above reappears. While the ensemble average follows Eq.(3.30), $\langle x^2(t) \rangle \propto t^{\alpha\gamma}$, the double time and ensemble average reveals a different behavior.

To calculate the double average $\left\langle x^2(n) \right\rangle \propto \langle \langle (n(t+\Delta t) - n(t))^\gamma \rangle_{ens} \rangle_T$ we proceed as follows. For given t the ensemble average corresponds to the average of $\langle n^\gamma(\Delta t; t_a) \rangle$ of the corresponding power of the number of steps of CTRW aged up to the time t_a. In all our previous considerations we had $\gamma = 1$, and this linearity was used extensively.

Let us denote by t_w the forward waiting time for the first step of a CTRW aged up to the time t_a. The number of steps n entering the average $\langle n^\gamma(\Delta t; t_a) \rangle_{ens}$ is the number of steps done by CTRW during the time $t' = \Delta t - t_w$ between the first step in the interval between t_a and $t_a + \Delta t$ (provided that a step was done in this interval). Therefore

$$\langle n^\gamma(\Delta t; t_a) \rangle_{ens} = \int_0^{\Delta t} \langle n^\gamma(\Delta t - t_w) \rangle_{ens} \psi_1(t_w; t_a) dt_w, \tag{4.15}$$

where $\psi_1(t_w;\ t_a)$ is given by Eq.(4.8). The ensemble mean $\langle n^\gamma(t) \rangle$ can be obtained using the methods described in Chapter 3,

$$\langle n^\gamma(t) \rangle \approx \int_0^\infty \frac{t}{\alpha\tau} n^{-\frac{1}{\alpha}-1+\gamma} L_\alpha \left(\frac{t}{\tau\, n^{1/\alpha}} \right) dn = \left(\frac{t}{\tau} \right)^{\alpha\gamma} \frac{\Gamma(1-\gamma)}{\Gamma(1-\alpha\gamma)}$$

(see the discussion of Eq.(3.26)). Now we perform integration over t_w in Eq.(4.15) and note that, since the behavior only for $\Delta t << t_a$ will be relevant for what follows, we can use the approximation $\psi_1(t_w;\ t_a) = \frac{\sin(\pi\alpha)}{\pi}\frac{t_a^{\alpha-1}}{t_w^{\alpha}}$. Thus

$$\langle n^\gamma(\Delta t; t_a)\rangle_{ens} \approx \frac{\sin(\pi\alpha)}{\pi}\frac{\Gamma(1-\gamma)}{\Gamma(1-\alpha\gamma)}\frac{t_a^{\alpha-1}}{\tau^{\alpha\gamma}}\int_0^{\Delta t}(\Delta t - t_w)^{\alpha\gamma}t_w{}^{-\alpha}dt_w.$$

The corresponding integral is the convolution of two power functions, which can be evaluated by means of Laplace transform and by applying Tauberian theorems:

$$\langle n^\gamma(\Delta t; t_a)\rangle_{ens} \approx const\cdot\frac{t_a^{\alpha-1}}{\tau^{\alpha\gamma}}\Delta t^{1-\alpha(1-\gamma)}.$$

Note that for $\gamma = 1$ we recover

$$\langle n^\gamma(\Delta t; t)\rangle_{ens} \approx const\cdot\frac{t^{\alpha-1}}{\tau^{\alpha}}\Delta t^1$$

as we should. The last step is the temporal averaging, which involves only t_a:

$$\langle\langle(n(t+\Delta t)-n(t))^\gamma\rangle_{ens}\rangle_T = \frac{1}{T}\int_0^T\langle n^\gamma(\Delta t; t_a)\rangle_{ens}\,dt_a \approx$$

$$const\cdot\frac{\Delta t^{1-\alpha(1-\gamma)}}{\tau^{\alpha\gamma}}\frac{1}{T}\int_0^T t_a^{\alpha-1}dt_a \propto \frac{\Delta t^{1-\alpha(1-\gamma)}}{\tau^{\alpha\gamma}T^{1-\alpha}}.$$

The MSD follows the behavior of the mean number of steps.

4.5 Response to the time-dependent field

Let us now consider another aspect of aging, namely that seen in the linear response of the system to an external field [4]. This one again can be dealt with on the basis of the discussion of the behavior of the mean number of steps.

Let us consider the response of the system to time-dependent field $f(t)$ (for the sake of simplicity we concentrate on the one-dimensional situation). We consider here a spatially homogeneous situation in an infinite system without boundaries. We assume that the field does not affect waiting times but biases the direction of the jump when the jump is performed.[1] The mean displacement during a jump taking place at time t is proportional to the instantaneous value of the field, $\overline{\Delta x} = \mu f(t)$. Let us now consider the ensemble average displacement $d\bar{x}$ between the time instants t and $t + dt$. Choosing small dt, we can see that $k(t)dt = \frac{d\langle n(t)\rangle}{dt}dt$ is exactly the probability that

[1] This is indeed true, e.g., for the comb model, if the acting force is parallel to the direction of the backbone. In this case it does not affect the motion in the teeth and hence the waiting times. In the case of a trap model (Exercise 3.9) this is true in the first order in the external force.

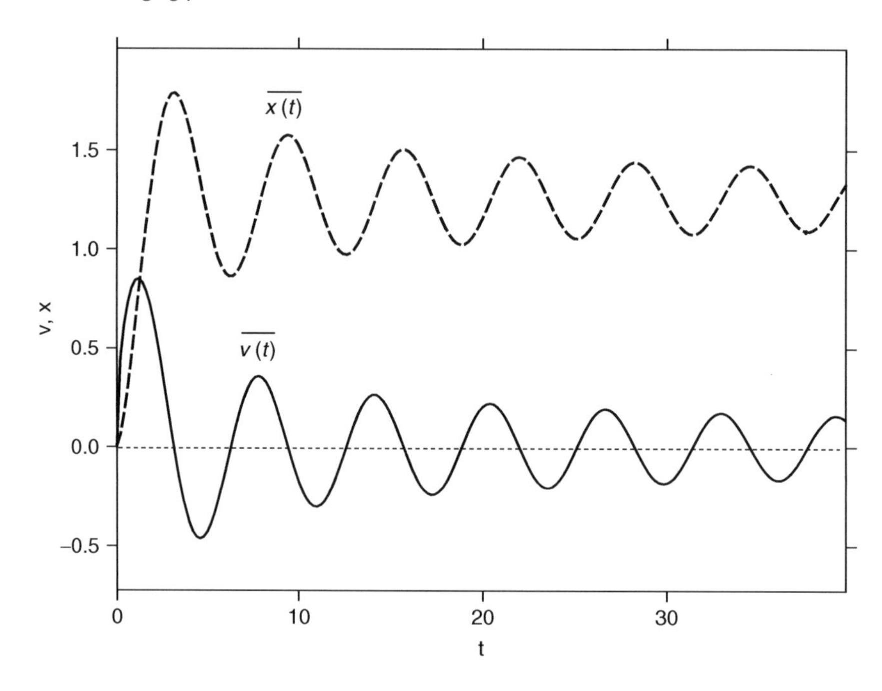

Fig. 4.3 *This figure shows (in arbitrary units) the response of an aging CTRW to a sinusoidal external field. While $\overline{v(t)}$ exhibits decaying oscillations around zero ("death of linear response"), $\overline{x(t)}$ attains a constant value, mostly acquired when the system was young ("Freudian memory").*

a step was made between t and $t + dt$, or a portion of realizations in which such a step was made (to see this, imagine you chose dt so small that during this time either no or only one step can take place in each realization). Therefore the mean displacement during the interval dt is $d\bar{x} = \mu f(t)k(t)dt$. This gives us the expression of the mean velocity $\bar{v}(t) = \mu f(t)k(t)$ at time t. Looking at the effective mobility connecting the instantaneous values of the field and of the mean velocity, we see that it is proportional to $\mu k(t)$. For CTRW with power-law waiting-time distribution $\psi(t) \propto t^{-1-\alpha}$ this decays as $t^{\alpha-1}$, i.e., the velocity as the response to the same field gets smaller over time. This effect is called "the death of linear response" in Ref. [5].

Integrating the mean velocity, we get the overall displacement $\overline{x(t)} = \int_0^t \mu f(t')k(t')dt'$. Since $k(t)$ decays over time, the displacement is mostly governed by the values of the field at early times, close to the time $t = 0$ when the system was prepared, and is much less sensitive to what happened later on, when the system was aged. This effect is called the "Freudian memory" in aging systems. An example of the corresponding behavior of $\overline{v(t)}$ and $\overline{x(t)}$ is shown in Fig. 4.3.

References

[1] M. Abramovitz and I.A. Stegun. *Handbook of Mathematical Functions*, New York: Dover, 1972

[2] F. Oberhettinger and L. Badii. *Tables of Laplace Transforms*, Berlin: Springer, 1973
[3] A. Lubelski, I.M. Sokolov, and J. Klafter. *Phys. Rev. Lett.* **100**, 250602 (2008)
[4] I.M. Sokolov, A. Blumen, and J. Klafter. *Physica A* **302**, 268 (2001)
[5] I.M. Sokolov and J. Klafter. *Phys. Rev. Lett.* **97**, 140602 (2006)

Further reading

W. Feller. *An Introduction to Probability Theory and Its Applications*, New York: Wiley, 1971 (Corresponding material is contained in Vol. 2, Ch. VI and Ch. XIV.)
D.R. Cox. *Renewal Theory*, New York: Wiley, 1962

5

Master equations

"All changes in nature are such that inasmuch is taken from one place insomuch is added to another."

Mikhail Lomonosov

Our probabilistic approach to random walks, put forward in Chapter 1 and pursued since then, immediately led to the results for the PDFs of the walker's positions at time t. Knowing this PDF, we were able to obtain many other results, such as first-passage or return probabilities. However, our approach relied on assumptions of spatial homogeneity of the system and (except for the discussion in Sec. 4.5) on the time-independence of step length distribution. Therefore this approach was appropriate for a description of homogeneous systems under no external forcing (or at best under a force that is independent of time and of coordinates). The assumption of the homogeneity is space and in time allowed us to turn to a Fourier–Laplace representation, where the integral recurrence equations for CTRWs took the form of simple algebraic ones. Although such homogeneous situations are widespread, they do not exhaust the whole variety of cases. Developing an instrument for describing the cases in which the assumption of special homogeneity no longer holds is exactly the aim of the present chapter. Our present approach is based on the master equations, the balance equations for the probabilities of finding a particle at some place (at a site $\mathbf{r}$ in a lattice walk or in a small volume dv around the point $\mathbf{r}$ in the continuous ones). The picture adopted here takes a different perspective from that of previous chapters: While there we were following the motion of the walker and interpreted $P(\mathbf{r}, t)$ as the probability of finding a walker at site $\mathbf{r}$ at time t, here we concentrate our attention on a site and consider $P(\mathbf{r}, t)$ as the probability that it is occupied by the walker.

The standard form of the master equation for normal diffusion on a lattice is [1]:

$$\frac{d}{dt}P(\mathbf{r},t) = \frac{1}{\tau}\sum_{\mathbf{r}'}[p(\mathbf{r},\mathbf{r}')P(\mathbf{r}',t) - p(\mathbf{r}',\mathbf{r})P(\mathbf{r},t)], \tag{5.1}$$

where $\frac{p(\mathbf{r},\mathbf{r}')}{\tau}$ is the rate of jumps from $\mathbf{r}'$ to $\mathbf{r}$. Owing to the balance of the probabilities, $\sum_{\mathbf{r}'} p(\mathbf{r}',\mathbf{r}) = 1$, Eq.(5.1) reduces to

$$\frac{d}{dt}P(\mathbf{r},t) = -\frac{1}{\tau}P(\mathbf{r},t) + \frac{1}{\tau}\sum_{\mathbf{r}'} p(\mathbf{r},\mathbf{r}')P(\mathbf{r},t). \tag{5.2}$$

As we go on to show, Eq.(5.1) corresponds to the case of exponential waiting-time densities. In the case of $\psi(t)$ different from an exponential, the corresponding master equation takes the form

$$\frac{d}{dt}P(\mathbf{r},t) = \int_0^t dt' \sum_{\mathbf{r}'} [p(\mathbf{r},\mathbf{r}',t-t')P(\mathbf{r}',t') - p(\mathbf{r}',\mathbf{r},t-t')P(\mathbf{r},t')]. \qquad (5.3)$$

This integro-differential equation is referred to as a *generalized master equation* (GME) [1]. In many cases, such as that of decoupled CTRW described in Chapter 3, the transition probability rates $p(\mathbf{r},\mathbf{r}',t)$ decouple into a product of a coordinate-dependent and a time-dependent function, $p(\mathbf{r},\mathbf{r}',t) = \phi(t)p(\mathbf{r},\mathbf{r}')$. Equation (5.1) corresponds to just such a form with $\phi(t) = \frac{1}{\tau}\delta(t)$ proportional to a δ-function.

In the next few paragraphs we first discuss under which conditions the GME describes CTRW, and then derive the GME immediately from the CTRW scheme.

Let us bring together some important results on CTRW on lattices, combining the discussions in Chapters 2 and 3, as summarized by Eq.(3.14) giving the relation between the Laplace-transformed probability of being at site $\mathbf{r}$ in CTRW and the generating function of simple random walks:

$$P(\mathbf{r},s) = \frac{1-\psi(s)}{s} P(\mathbf{r}; z = \psi(s)).$$

Equation (5.3) for decoupled transition rates can be rewritten as

$$\frac{d}{dt}P(\mathbf{r},t) = \int_0^t dt' \phi(t-t') \left[-P(\mathbf{r},t') + \sum_{\mathbf{r}'} p(\mathbf{r},\mathbf{r}')P(\mathbf{r}',t') \right]. \qquad (5.4)$$

Taking a Laplace transform of this equation (having the structure of a convolution), we obtain

$$sP(\mathbf{r},s) - P(\mathbf{r},t=0) = \phi(s) \left[-P(\mathbf{r},s) + \sum_{\mathbf{r}'} p(\mathbf{r},\mathbf{r}')P(\mathbf{r}',s) \right] \qquad (5.5)$$

where $P(\mathbf{r},t=0) = \delta_{\mathbf{r},0}$ is the initial condition. We now reorganize the terms in Eq.(5.5) to get

$$(s+\phi(s))P(\mathbf{r},s) - \delta_{\mathbf{r},0} = \frac{\phi(s)}{s+\phi(s)} \sum_{\mathbf{r}'} p(\mathbf{r},\mathbf{r}')[s+\phi(s)]P(\mathbf{r}',s). \qquad (5.6)$$

Let us also recall that in the case of translationally invariant $p(\mathbf{r},\mathbf{r}') = p(\mathbf{r}-\mathbf{r}')$ the generating function $P(\mathbf{r},z)$ fulfills Eq.(2.7),

$$P(\mathbf{r},z) - z\sum_{\mathbf{r}'} p(\mathbf{r}-\mathbf{r}')P(\mathbf{r}',z) = \delta_{\mathbf{r},0},$$

which we now compare to Eq.(5.6). Specifically, we identify z with $\psi(s)$, which then is defined as $z = \psi(s) = \frac{\phi(s)}{s+\phi(s)}$, which establishes the equivalence between CTRW and GME. Inverting this relation, we can express $\phi(s)$ through $\psi(s)$ via $\phi(s) = s\frac{\psi(s)}{1-\psi(s)}$. The function

$$M(s) = \frac{\psi(s)}{1 - \psi(s)} \tag{5.7}$$

is connected with the rate of steps discussed in Chapter 4 and will repeatedly appear in what follows.

Exercise 5.1 Show that CTRWs with $\psi(t) = \frac{1}{\tau}e^{-t/\tau}$ are described by the GME with $\phi(t) = \frac{1}{\tau}\delta(t)$, i.e., by an ordinary master equation, Eq.(5.1).

In what follows, the kernel $\phi(t)$ in the GME, Eq.(5.5), will be called a memory kernel, since Eq.(5.5) connects the evolution of the PDF of the walker's positions at time t with those at all previous times. The CTRW process, described by this equation, is, however, practically memoryless: The sojourn durations and step lengths in different jumps are independent; the process is Markovian (i.e., memoryless) when considered to be dependent on the number of steps (internal time) n, but whenever the waiting times are non-exponential some special type of "memory" appears (as exemplified by the inspection paradox discussed in Chapter 4). Such processes are called *renewal* or *semi-Markov* processes.

5.1 A heuristic derivation of the generalized master equation

In order to make the approach clear and to show how it corresponds to those used in the previous chapters, we start here with the same decoupled random walks homogeneous in space and in time and give a heuristic derivation of the master equation for this case (see, for example, Ref.[2] and references therein). The waiting-time PDF $\psi(t)$ are assumed to be the same for all sites. The transition probabilities are taken to depend only on the distance between the sites $p(\mathbf{r},\mathbf{r}') = p(\mathbf{r} - \mathbf{r}')$, but not on time. These are exactly the assumptions made in our discussion above.

Let us consider a walker sitting at a site $\mathbf{r}$ and about to perform a step between time t and time $t + dt$ (see Fig. 5.1). The master equations are the equations for the balance of probability for the particle's jumps between the sites. Let $P(\mathbf{r}, t)$ be the probability of finding a particle at site $\mathbf{r}$ at time t. The balance equation for the state $\mathbf{r}$ reads

$$\frac{dP(\mathbf{r}, t)}{dt} = j^+(\mathbf{r}, t) - j^-(\mathbf{r}, t), \tag{5.8}$$

with $j^\pm(\mathbf{r}, t)$ denoting the gain and loss fluxes (i.e., the probability gain and loss at site $\mathbf{r}$ per unit time).

Now, let us consider a particle leaving the site $\mathbf{r}$ in a time interval between t and $t + dt$ (with probability $j^-(\mathbf{r}, t)dt$). This particle was either at $\mathbf{r}$ from the very beginning

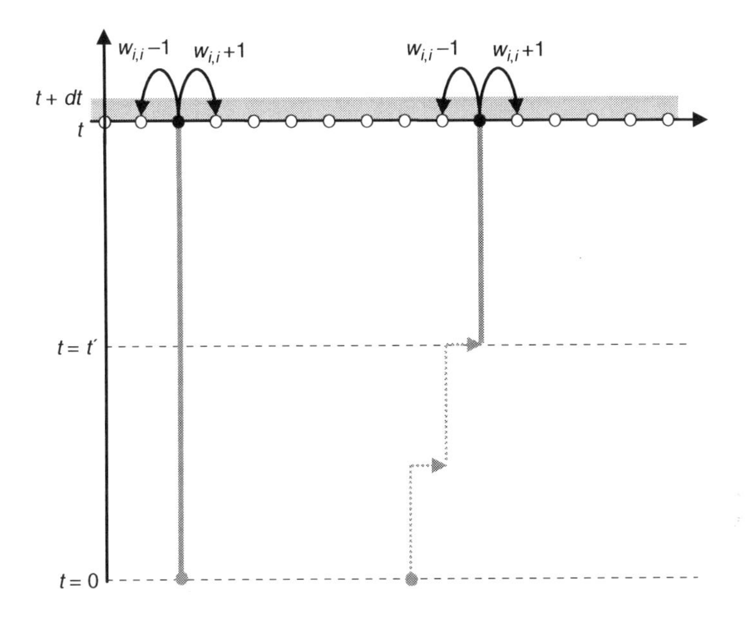

Fig. 5.1 *As an explanation of our approach, we show here in one dimension the trajectories of the two particles about to perform a step between t and $t+dt$: The left particle has been at the same site since the time the system was created; the right particle started from a different site and arrived at its actual position at time $0 < t' < t$.*

(i.e., from the time $t = 0$) or it arrived at $\mathbf{r}$ at some later instant of time t', $0 < t' < t$. In the former case the probability of leaving between t and $t + dt$ is $\psi(t)dt$, and the contribution of such particles to the overall probability loss is $\psi(t)P(\mathbf{r}, 0)dt$, where $P(\mathbf{r}, 0)$ gives the probability that the particle was at $\mathbf{r}$ at time $t = 0$. In the latter case, it is of no importance where the particle came from, since the memory on previous steps is erased upon arrival (a semi-Markovian property), and the probability of leaving the site between t and $t + dt$ is $\psi(t - t')dt$. Since the probability of arriving at a site between t' and $t' + dt'$ is given by $j^+(\mathbf{r}, t')dt'$, the loss flux at time t, given by the integration over the arrival times t', reads

$$j^-(\mathbf{r},t) = \psi(t)P(\mathbf{r},0) + \int_0^t \psi(t-t')j^+ \int_0^t \psi(t-t')j^+(\mathbf{r},t')\,dt'.$$

Expressing $j^+(\mathbf{r}, t)$ through $j^-(\mathbf{r}, t)$ and $P(\mathbf{r}, t)$ with the help of Eq.(5.8), we get an integral equation connecting $j^-(\mathbf{r}, t)$ and $P(\mathbf{r}, t)$ at the same site $\mathbf{r}$:

$$j^-(\mathbf{r},t) = \psi(t)P(\mathbf{r},0) + \int_0^t \psi(t-t')\left[\frac{dP(\mathbf{r},t')}{dt'} + j^-(\mathbf{r},t')\right] dt'. \tag{5.9}$$

Equation (5.9) allows us to express $j^-(\mathbf{r},t)$ through $P(\mathbf{r},t)$. Turning to the Laplace transform we note that $L\left\{\frac{dP(\mathbf{r},t')}{dt'}\right\} = s\tilde{P}(\mathbf{r},s) - P(\mathbf{r},t=0)$. Here we use the tilde ($\sim$) to denote the Laplace-transformed $\tilde{P}(\mathbf{r},s)$ of $P(\mathbf{r},t)$, since it falls within the same equation as the non-transformed initial condition $P(\mathbf{r},0)$. We now get:

$$j^-(\mathbf{r},s) = \psi(s)P(\mathbf{r},0) + \psi(s)\left[u\tilde{P}(\mathbf{r},s) - P(\mathbf{r},0) + j^-(\mathbf{r},s)\right],$$

from which we readily obtain

$$j^-(\mathbf{r},s) = s\frac{\psi(s)}{1-\psi(s)}\tilde{P}(\mathbf{r},s) = sM(s)\tilde{P}(\mathbf{r},s).$$

with $M(s)$ given by Eq.(5.7). The above equation connects $j^-(\mathbf{r},t)$ and $P(\mathbf{r},t)$ at the same site and no longer contains the initial condition $P(\mathbf{r},0)$ explicitly. In the time domain, $j^-(\mathbf{r},t)$ can be expressed through an integral operator $\hat{\Phi}$ acting on the function of time $P(\mathbf{r},t)$,

$$j^-(\mathbf{r},t) = \hat{\Phi}P(\mathbf{r},t) = \frac{d}{dt}\int_0^t M(t-t')P(\mathbf{r},t')dt', \tag{5.10}$$

with the integral kernel $M(t)$ being the inverse Laplace transform of $M(u)$. This integral kernel $M(t)$ has a clear physical meaning. Noting that

$$M(s) = \frac{\psi(s)}{1-\psi(s)} = \psi(s) + \psi^2(s) + \ldots + \psi^n(s) + \ldots$$

Exercise 5.2 Find the form of the memory kernel $M(t)$ for the case of the exponential waiting-time probability density $\psi(t) = \tau^{-1}\exp(-t/\tau)$.

Exercise 5.3 Find the asymptotical form of the memory kernel $M(t)$ for large t in the case of the power-law waiting-time probability density $\psi(t) \propto t^{-1-\alpha}$.

we see that it corresponds to the sum of multiple convolutions of $\psi(t)$ with itself: $M(t) = \psi(t) + \psi(t)^*\psi(t) + \ldots + \psi(t)^*\psi(t)^* \ldots^* \psi(t) + \ldots$. Therefore it represents the time-dependent density of steps: $M(t)dt$ is the mean number of steps performed by the particle within a time window $[t, t+dt]$, and differs from the step rate $k(t)$ only with respect to the fact that the zeroth step (at $t=0$) is not counted.

Let us now turn to the probability conservation for jumps between different sites. This condition connects the gain current at a site k with loss currents at its neighbors:

$$j^+(\mathbf{r},t) = \sum_{\mathbf{r}'} p(\mathbf{r},\mathbf{r}')j^-(\mathbf{r}',t), \tag{5.11}$$

where the sum runs over all sites from which the transitions to site $\mathbf{r}$ are possible. As before, $\sum_{\mathbf{r}'} p(\mathbf{r},\mathbf{r}') = 1$. We now insert Eq.(5.11) into Eq.(5.8) and get

$$\frac{dP(\mathbf{r},t)}{dt} = \sum_{\mathbf{r}'} p(\mathbf{r},\mathbf{r}') j^-(\mathbf{r}',t) - j^-(\mathbf{r},t). \tag{5.12}$$

Using Eq.(5.10) connecting loss fluxes with the probabilities $P(\mathbf{r},t)$, we obtain the final result

$$\frac{dP(\mathbf{r},t)}{dt} = \sum_{\mathbf{r}'} p(\mathbf{r},\mathbf{r}') \frac{d}{dt} \int_0^t M(t-t')P(\mathbf{r}',t')dt' - \frac{d}{dt} \int_0^t M(t-t')P(\mathbf{r},t')dt', \tag{5.13}$$

representing a GME. The GME with the constant memory kernel $M(t) = 1/\tau$ (corresponding to the exponential waiting-time distribution; see Exercise 5.1) takes an especially simple form:

$$\frac{dP(\mathbf{r},t)}{dt} = \frac{1}{\tau} \left[\sum_{\mathbf{r}} p(\mathbf{r},\mathbf{r}')P(\mathbf{r}',t') - P(\mathbf{r},t') \right], \tag{5.14}$$

which is a customary (Pauli) master equation.

For the time-independent transition probabilities $p(\mathbf{r},\mathbf{r}')$ we can interchange the sequence of summation over sites and integration over t to put the GME into a simpler form

$$\frac{dP(\mathbf{r},t)}{dt} = \frac{d}{dt} \int_0^t dt' M(t-t') \left[\sum_{\mathbf{r}'} p(\mathbf{r},\mathbf{r}')P(\mathbf{r}',t') - P(\mathbf{r},t') \right]. \tag{5.15}$$

Note that Eq.(5.15) is equivalent to Eq.(5.4) if we take $\phi(t) = \frac{d}{dt}M(t) + M(0)\delta(t)$, which can be immediately proved by substitution.

5.1.1 Master equations for off-lattice walks

We can dispense with the assumption of the existence of the lattice and obtain the balance equations for a more general case. Instead of considering lattice sites, we now subdivide the space into compartments of size (volume) dv (being dx in one dimension, $dxdy$ in two dimensions, and $dxdydz$ in three dimensions) and give the probability of the particle being in the corresponding compartment by denoting it as $P(\mathbf{r},t)dv$. We note that the on-site balance equation

$$\frac{dP(\mathbf{r},t)}{dt} = j^+(\mathbf{r},t) - j^{-1}(\mathbf{r},t)$$

as well as its direct consequence

$$j^-(\mathbf{r},t) = \frac{d}{dt} \int_0^t M(t-t')P(\mathbf{r},t')dt'$$

stay valid also for the off-lattice situation. Assuming $p(\mathbf{r},\mathbf{r}')dvdv'$ to be the transition probability between the two corresponding compartments (in the homogeneous case, this depends only on the jump distance), we get for the gain flux

$$j^+(\mathbf{r},t) = \int_{\Omega} j^-(\mathbf{r}',t)\lambda(\mathbf{r}-\mathbf{r}')d\mathbf{r}',$$

where the integration is over the whole volume of the system (except for the infinitesimally small region dv around the point $\mathbf{r}$), so that the overall balance equation reads

$$\frac{dP(\mathbf{r},t)}{dt} = \frac{d}{dt}\int_0^t dt' \int_V M(t-t')p(\mathbf{r},\mathbf{r}')P(\mathbf{r}',t')d\mathbf{r}' - \frac{d}{dt}\int_0^t M(t-t')p(\mathbf{r},t')dt'$$

(where we have neglected the excluded infinitesimal volume). Now, assuming the waiting-time distribution to be independent of positions and the jump length distribution to be independent of times, we can rewrite this in the form

$$\frac{dP(\mathbf{r},t)}{dt} = \frac{d}{dt}\int_0^t dt' M(t-t') \int_V P(\mathbf{r}',t')[p(\mathbf{r},\mathbf{r}') - \delta(\mathbf{r}-\mathbf{r}')]d\mathbf{r}', \tag{5.16}$$

where we have included the second term (not containing the volume integral) into the first one with the help of the δ-function. In the case where the transition probability depends only on the distance $\mathbf{r}-\mathbf{r}'$ the equation reads:

$$\frac{dP(\mathbf{r},t)}{dt} = \frac{d}{dt}\int_0^t dt' M(t-t') \int_V P(\mathbf{r}',t')\left[\lambda(\mathbf{r}-\mathbf{r}') - \delta(\mathbf{r}-\mathbf{r}')\right]d\mathbf{r}'. \tag{5.17}$$

Exercise 5.4 Prove that the solution to Eq.(5.17) in the Fourier–Laplace space reads $P(\mathbf{k},s) = \frac{1-\psi(s)}{s}\frac{1}{1-\psi(s)\lambda(\mathbf{k})}$, i.e., presents a propagator of a random walk.

5.2 A note on time-dependent transition probabilities

The master equation approach is versatile and can be generalized to many different situations such as that with time-dependent transition probabilities or with waiting times that change from one site to another. Let us return to our lattice model. If the external time-dependent field does not change the waiting-time distribution but biases only the jump direction (introduces changes in $p(\mathbf{r},\mathbf{r}',t)$), it affects only the balance equation Eq.(5.11) for transitions between the sites. On the other hand, when we consider, e.g., the position-dependent waiting-time distributions $\psi_{\mathbf{r}}(t)$ (which situation corresponds, e.g., to the subdiffusive medium with temperature gradient, or to the contact of media with different properties), these affect only the on-site balance equation, Eq.(5.10), leading to the fact that the memory kernel M now depends on the site position $\mathbf{r}$. As long as transition probabilities do not depend on t explicitly, the summation (or integration) over sites can be interchanged with the temporal integro-differential operator with kernel $M_{\mathbf{r}}(t) = M(\mathbf{r}, t)$, as follows from the on-site balance. In the case where the transition probabilities depend on time explicitly, such an interchange in the sequences of integration in time and summation over sites becomes impossible, and the master equation in the following form

$$\frac{dP(\mathbf{r},t)}{dt} = \sum_{\mathbf{r}'} p(\mathbf{r},\mathbf{r}',t)\frac{d}{dt}\int_0^t M_{\mathbf{r}'},(t-t')P(\mathbf{r}',t')dt' - \frac{d}{dt}\int_0^t M_{\mathbf{r}}(t-t')P(\mathbf{r},t')dt'$$

should be used. The corresponding form for the off-lattice walk reads:

$$\frac{dP(\mathbf{r},t)}{dt} = \int_V [p(\mathbf{r},\mathbf{r}',t) - \delta(\mathbf{r}-\mathbf{r}')] \left(\frac{d}{dt}\int_0^t dt' M(\mathbf{r}',t-t')P(\mathbf{r}',t') \right) d\mathbf{r}'. \tag{5.18}$$

5.3 Relation between the solutions to the generalized and the customary master equations

For time-independent transition probabilities, an intimate relation exists between the solution to the customary (Pauli) master equation and the GME. It does not matter whether we consider a lattice or an off-lattice situation.

Let us consider the two equations, Eq.(5.16) and its customary analog,

$$\frac{df(\mathbf{r},t)}{dt} = \hat{L}f(\mathbf{r},t) = \int_V f(\mathbf{r}',t)[p(\mathbf{r},\mathbf{r}') - \delta(\mathbf{r}-\mathbf{r}')]d\mathbf{r}' \tag{5.19}$$

with the same spatial linear operator $\hat{L}$ acting on the probability density on the right-hand side, which we now denote by $f(\mathbf{r},t)$ to distinguish it from the solution $P(\mathbf{r},t)$ to Eq.(5.16). Equation (5.19) corresponds exactly to Eq.(5.16) with $M(t) = \delta(t)$. We moreover assume that both equations have the same initial conditions $f(\mathbf{r},0) = p(\mathbf{r},0)$ and the same boundary conditions. The solutions to Eqs.(5.16) and (5.19) are then connected to each other by the following integral transformation:

$$P(\mathbf{r},t) = \int_0^\infty f(\mathbf{r},\tau)T(\tau,t)d\tau, \tag{5.20}$$

with $T(\tau,t)$ defined by its Laplace transform in its second variable $T(\tau,s) = \int_0^\infty T(\tau,t)e^{-st}dt$:

$$T(\tau,s) = \frac{1}{sM(s)}\exp\left[-\frac{\tau}{M(s)}\right]. \tag{5.21}$$

To see this let us apply the Laplace transform to Eqs.(5.16) and (5.19) to get

$$s\tilde{f}(\mathbf{r},s) - P(\mathbf{r},0) = \hat{L}\tilde{f}(\mathbf{r},s) \tag{5.22}$$

and

$$s\tilde{P}(\mathbf{r},s) - P(\mathbf{r},0) = sM(s)\hat{L}\tilde{P}(\mathbf{r},s), \tag{5.23}$$

where we have used the assumption that $f(\mathbf{r},0) = P(\mathbf{r},0)$. Here the Laplace-transformed functions are denoted by a tilde ($\sim$), since they appear in the same equations as the originals. Now we substitute our integral solution, Eq.(5.20), into Eq.(5.19). The Laplace transform of $p(\mathbf{r},t)$ reads:

$$\begin{aligned} P(\mathbf{r},s) &= \int_0^\infty dt e^{-st} \int_0^\infty d\tau f(\mathbf{r},\tau) T(\tau,t) \\ &= \int_0^\infty d\tau \frac{1}{sM(s)} \exp\left[-\frac{\tau}{M(s)}\right] f(\mathbf{r},\tau) \\ &= \frac{1}{sM(s)} \tilde{f}\left[\mathbf{r}, \frac{1}{M(s)}\right] \end{aligned} \tag{5.24}$$

so that Eq.(5.23) takes the form

$$\frac{1}{M(s)} = \tilde{f}\left[\mathbf{r}, \frac{1}{M(s)}\right] - P(\mathbf{r},0) = \hat{L}\tilde{f}\left[\mathbf{r}, \frac{1}{M(s)}\right].$$

Changing the variable to $u = 1/M(s)$, we recover Eq.(5.22), proving that the connection assumed is indeed valid.

The usefulness of the connection discussed is manifold. First, in many cases the solutions to the ordinary master equation are known, and Eqs.(5.20) and (5.21) allow us to obtain the solution for a generalized equation without explicitly solving it. Second, it is a useful theoretical instrument for proving the existence (i.e., the non-negativity) of the solution in many cases, where it cannot be written down explicitly. This is particularly the case when $T(\tau,t)$ can be proved to be a probability density in τ. To prove this, it is enough to prove the non-negativity of $T(\tau, t)$ since its normalization is guaranteed by construction: Using Eq.(5.21) we find that the Laplace transform of $\int_0^\infty T(\tau,t)dt$ in its second variable reads

$$\int_0^\infty T(\tau,s)d\tau = \int_0^\infty \frac{d\tau}{sM(s)} \exp\left[-\frac{\tau}{M(s)}\right] = \frac{1}{s},$$

which corresponds to $\int_0^\infty T(\tau,t)dt = 1$.

Provided $T(\tau, t)$ is non-negative, this can be interpreted as a probability density of the variable τ at time t, and the integral expression, Eq.(5.20), can then be interpreted as an integral analog of Eq.(3.8), and corresponds to the so-called integral formula of subordination. Using the mathematical terminology of subordinated processes [3], Eq.(5.19) can then be considered as the master equation for the parent process, and $T(\tau, t)$ is the probability density for the directing process $\tau(t)$ at time t (however, this is not enough to show that the random process $\mathbf{r}(t)$ is really subordinated to $\mathbf{r}(\tau)$, since we have not shown that $\tau(t)$ has non-negative increments, which is necessary if we want to interpret τ as an operational time).

5.4 Generalized Fokker–Planck and diffusion equations

Looking at the behavior of the random walks only at large scales, i.e., those much larger than the typical step's size $\langle(\mathbf{r}-\mathbf{r}')^2\rangle^{1/2}$, we can often simplify the situation by passing from the master equation, Eq.(5.16), which is (in general) integral in its spatial coordinates

$$\frac{dP(\mathbf{r},t)}{dt} = \frac{d}{dt}\int_0^t dt' M(t-t') \int_V P(\mathbf{r}',t')[p(\mathbf{r},\mathbf{r}') - \delta(\mathbf{r}-\mathbf{r}')]d\mathbf{r}'$$

to a differential one. To do so let us expand the functions $P(\mathbf{r}, t)$ into Taylor series around $\mathbf{r} = \mathbf{r}'$:

$$P(\mathbf{r}',t) \cong P(\mathbf{r},t) + \nabla P(\mathbf{r},t)(\mathbf{r}'-\mathbf{r}) + \frac{1}{2}\Delta P(\mathbf{r},t)(\mathbf{r}'-\mathbf{r})^2 + \ldots$$

Inserting such an expansion into Eq.(5.16) leads to a partial differential equation for $P(\mathbf{r}, t)$ that has the form

$$\frac{dP(\mathbf{r},t)}{dt} = \frac{d}{dt}\int_0^t dt' M(t-t') \left[\mathbf{A}(\mathbf{r})\nabla P(\mathbf{r},t) + \frac{B(\mathbf{r})}{2}\Delta P(\mathbf{r},t)\right] \tag{5.25}$$

with the coefficients $\mathbf{A}(\mathbf{r}) = \int_V p(\mathbf{r},\mathbf{r}')(\mathbf{r}-\mathbf{r}')d\mathbf{r}'$ and $B(\mathbf{r}) = \int_V p(\mathbf{r},\mathbf{r}')(\mathbf{r}-\mathbf{r}')^2 d\mathbf{r}'$. Equation (5.25) is the *generalized Fokker–Planck equation* (a customary Fokker–Planck equation is that with $M(t) = \delta(t)$). For homogeneous systems the coefficients are constant, i.e., independent of $\mathbf{r}$.

Parallel to the case of a customary Fokker–Planck equation, taking into account more than two first terms in the expansion of $P(\mathbf{r}, t)$ may lead to oscillating (i.e., no longer non-negative definite) densities. Thus, if the accuracy of the approximation, Eq.(5.25), is not sufficient, the whole integral form of the master equation should be used.

Exercise 5.5 Consider the generalized Fokker–Planck equation for symmetric, homogeneous CTRW, Eq.(5.25) with $\mathbf{A}(\mathbf{r}) = 0$ and $B(\mathbf{r}) = const.$ Prove that the solution to Eq.(5.25) in the Fourier–Laplace space corresponds to the small-k limit of the solution to Eq.(5.16), see Exercise 5.4.

For random walks describing systems possessing true thermodynamic equilibrium a different notation is often used. In such systems the equilibrium probability of finding a particle (walker) at position $\mathbf{r}$ is proportional to $P_{eq}(\mathbf{r}) \propto \exp\left(-\frac{U(\mathbf{r})}{kT}\right)$, with $U(\mathbf{r})$ being the potential energy of the walker at position $\mathbf{r}$, as follows from the Boltzmann distribution. The transition probabilities moreover fulfill the detailed balance condition $P_{eq}(\mathbf{r}')p(\mathbf{r},\mathbf{r}') = P_{eq}(\mathbf{r})p(\mathbf{r}',\mathbf{r})$ or

$$\frac{p(\mathbf{r},\mathbf{r}')}{p(\mathbf{r}'\mathbf{r})} = \exp\left(\frac{U(\mathbf{r}') - U(\mathbf{r})}{KT}\right) \tag{5.26}$$

stating that during any time interval the mean number of transitions between two arbitrary sites in equilibrium is equal in both directions. The detailed balance condition is the consequence of the second law of thermodynamics. The transition probability between the two sites $\mathbf{r}'$ and $\mathbf{r}$ depends therefore on the potential difference between these sites: $p(\mathbf{r},\mathbf{r}') = p(\mathbf{r},\mathbf{r}'; V(\mathbf{r},\mathbf{r}'))$, with $V(\mathbf{r},\mathbf{r}') = U(\mathbf{r}') - U(\mathbf{r})$. Let us introduce the potential force $\mathbf{f}(\mathbf{r})$ corresponding to $U(\mathbf{r})$, $\mathbf{f}(\mathbf{r}) = -\nabla U(\mathbf{r})$. Let us moreover assume that in the absence of the force $\mathbf{f}(\mathbf{r})$ the transition probability depends only on the difference of coordinates, $p(\mathbf{r},\mathbf{r}'; 0) = p(\mathbf{r} - \mathbf{r}')$ (i.e., the system in the absence of the force is homogeneous). This means that in the absence of the force the jumps are symmetric, since according to Eq.(5.26) $p(\mathbf{r},\mathbf{r}') = p(\mathbf{r}',\mathbf{r})$ and therefore $p(\mathbf{r} - \mathbf{r}') = p(\mathbf{r}' - \mathbf{r})$. Thus $p(\mathbf{r},\mathbf{r}'; 0)$ possesses zero first moment, $\int_V p(\mathbf{r},\mathbf{r}'; 0)(\mathbf{r}\text{-}\mathbf{r}')d\mathbf{r}' = 0$. Switching on the force $\mathbf{f}(\mathbf{r})$ introduces the bias in the distribution of step lengths and directions. For small $\mathbf{f}(\mathbf{r})$ the following expansion can be therefore used:

$$\begin{aligned} p(\mathbf{r},\mathbf{r}'; V) &= p(\mathbf{r},\mathbf{r}'; 0) + \frac{\partial}{\partial V} p(\mathbf{r},\mathbf{r}'; V)(U(\mathbf{r}) - U(\mathbf{r}')) + \dots \\ &= p(\mathbf{r},\mathbf{r}', 0) - \frac{\partial}{\partial V} p(\mathbf{r},\mathbf{r}'; V)\Big|_{V=0} \mathbf{f}(\mathbf{r})(\mathbf{r} - \mathbf{r}') \\ &\quad - \frac{1}{2}\frac{\partial}{\partial V} p(\mathbf{r},\mathbf{r}'; V)\Big|_{V=0} \nabla\mathbf{f}(\mathbf{r})(\mathbf{r} - \mathbf{r}')^2 \dots \end{aligned} \tag{5.27}$$

where in the second line we have used the approximation for the potential difference:

$$\begin{aligned} U(r) - U(r') &= \int_{\mathbf{r}'}^{\mathbf{r}} \mathbf{f}(\mathbf{x})d\mathbf{x} \approx -\frac{1}{2}\left[\mathbf{f}(\mathbf{r}) + \mathbf{f}(\mathbf{r}')\right](\mathbf{r} - \mathbf{r}') \\ &= -\mathbf{f}(\mathbf{r})(\mathbf{r} - \mathbf{r}') - \frac{1}{2}\nabla\mathbf{f}(\mathbf{r})(\mathbf{r} - \mathbf{r}')^2 + \dots. \end{aligned}$$

Using the form of the transition probability, Eq.(5.27), we get in the second order in $(\mathbf{r}-\mathbf{r}')$:

$$\begin{aligned}\frac{\partial P(\mathbf{r},t)}{\partial t} &= \frac{\partial}{\partial t}\int_0^t dt' M(t-t')\left[-\mu\mathbf{f}(\mathbf{r})\nabla P(\mathbf{r},t') - \mu P(\mathbf{r},t')\nabla\mathbf{f}(\mathbf{r}) + D\Delta P(\mathbf{r},t')\right] \\ &= \frac{\partial}{\partial t}\int_0^t dt' M(t-t')\nabla\left[-\mu\mathbf{f}(\mathbf{r})P(\mathbf{r},t') + D\nabla P(\mathbf{r},t')\right], \end{aligned} \tag{5.28}$$

with $D = \frac{1}{2}\int p(\mathbf{r}-\mathbf{r}')(\mathbf{r}-\mathbf{r}')^2 d\mathbf{r}'$ and $\mu = \int \frac{\partial}{\partial V} p(\mathbf{r},\mathbf{r}';V)\big|_{V=0}(\mathbf{r}-\mathbf{r}')^2 d\mathbf{r}'$. We note that the two constants are not independent.

Under the assumptions made, from the detailed balance it follows that $p(\mathbf{r},\mathbf{r}';V) = p(\mathbf{r}',\mathbf{r};-V)$. Differentiating both sides of Eq.(5.26) with respect to V, we get $\frac{\partial}{\partial V}\frac{p(\mathbf{r},\mathbf{r}';V)}{p(\mathbf{r},\mathbf{r}';-V)} = \frac{1}{kT}\exp\left(\frac{V}{kT}\right)$. Taking $V=0$ now gives $\frac{\partial}{\partial V}p(\mathbf{r},\mathbf{r}';V)\big|_{V=0} = \frac{1}{2kT}p(\mathbf{r},\mathbf{r}';V)$ so that

$$\mu = \frac{D}{kT}, \tag{5.29}$$

which is an expression called the generalized Einstein relation. The generalized Fokker–Planck equation, Eq.(5.28), and the corresponding Einstein relation, Eq.(5.29), will repeatedly appear in what follows.

References

[1] J.W. Haus and K.W. Kehr. *Phys. Repts.* **150**, 263 (1987)
[2] I.M. Sokolov and J. Klafter. *Chaos, Solitons and Fractals* **34**, 81 (2007)
[3] W. Feller. *An Introduction to Probability Theory and Its Applications*, New York: Wiley, 1971 (See Vol. 2, Ch. X.)

Further reading

N.G. van Kampen. *Stochastic Processes in Physics and Chemistry*, 3rd Edition, Amsterdam: North-Holland, 2007
W. Ebeling and I.M. Sokolov. *Statistical Thermodynamics and Stochastic Theory of Nonequilibrium Systems*, Singapore: World Scientific, 2005
B.D. Hughes. *Random Walks and Random Environments*, Vol. 1: *Random Walks*, Oxford: Clarendon, 1996
M. Magdziarz, A. Weron and J. Klafter, Phys. Rev. Lett., **101** , 210601 (2008)
A. Weron, M. Magdziarz and K. Weron, Phys. Rev. E, **77**, 036704 (2008)

6

Fractional diffusion and Fokker–Planck equations for subdiffusion

"God gave us the integers, all else is the work of man."

Leopold Kronecker

Let us return to the generalized Fokker–Planck equation considered in Chapter 5,

$$\frac{\partial p(\mathbf{r},t)}{\partial t} = \frac{\partial}{\partial t}\int_0^t dt' M(t-t')\nabla\left[-\mu \mathbf{f} p(\mathbf{r},t') + D\nabla p(\mathbf{r},t')\right],$$

and consider the case of the heavy-tailed waiting-time distribution $\psi(t) \cong t^{-1-\alpha}$. The memory kernel $M(t)$ in the generalized Fokker–Planck equation is given asymptotically (for large t) by $M(t) \cong t^{\alpha-1}$ (see Exercise 5.3). Thus the right-hand side of this equation contains an integro-differential operator

$$\frac{d}{dt}\int_0^t \frac{dt'}{(t-t')^{\alpha}} g(t')$$

acting on the function of time $g(t)$ (in this case the usual Fokker–Planck expression). This operator is proportional to that known in mathematics as a *Riemann–Liouville fractional derivative* and denoted ${}_0D_t^{1-\alpha}g(t)$. The main properties of this operator are discussed in Sec. 6.1.

6.1 Riemann–Liouville and Weyl derivatives

The question of how to generalize classical calculus to non-integer derivatives bothered mathematicians from the very beginning of the history of calculus itself. All modern definitions are based on various implementations of the following idea: Taking a derivative is a procedure that is inverse to integration $\int_a^t f(t')dt'$. The n-th derivative corresponds thus to an operation that is inverse to the n-fold repeated integration $\int_a^t dt_1 \int_a^{t_1} dt_2 \dots \int_a^{t_n} dt_n f(t_n)$. Crucial is the integral identity

$$\int_a^t \int_a^{t_1} \cdots \int_a^{t_{n-1}} f(t_n)dt_n \ldots dt_1 = \frac{1}{(n-1)!} \int_a^t (t-t')^{n-1} f(t')dt'. \tag{6.1}$$

We can now define a fractional integral from a to t of order α as

> **Exercise 6.1** Prove Eq.(6.1). One of the ways to do this is to use mathematical induction.

$${}_aI_t^\alpha f(t) = \frac{1}{\Gamma(\alpha)} \int_a^t (t-t')^{\alpha-1} f(t')dt', \tag{6.2}$$

which, for sufficiently regular functions, exists for all $\alpha > 0$. Note that $\Gamma(\alpha)$ is exactly the generalization of a factorial in the previous expression, since $\Gamma(n) = (n-1)!$. The fractional derivative of an arbitrary order β is defined by fractional integration and subsequent "normal" differentiation, namely

$${}_aD_t^\beta f(t) = \frac{d^n}{dt^n} {}_aI_t^{n-\beta}, \tag{6.3}$$

with $n = [\beta] + 1$, where $[\beta]$ denotes the whole part of β. Note that the order of the fractional integral $\alpha = n - \beta$ is between 0 and 1. Just like the corresponding fractional integral, the fractional derivative bears the dependence on the lower boundary of integration, which disappears only for whole values of β (see Exercise 6.2).

Equations (6.2) and (6.3) define the so-called Riemann–Liouville derivative, which possesses several properties in common with ordinary, whole-number derivatives. For example, it reproduces the standard results for differentiation of the power functions. Thus, the n-th derivative of the function x^m for $m \geq n$ reads:

$$\frac{d^n}{dx^n} x^m = m(m-1)\ldots(m-n+1)x^{m-n} = \frac{m!}{(m-n)!} x^{m-n}$$

and is equal to zero for $m < n$. Using the connection of factorials with the Gamma functions this can be rewritten as

$$\frac{d^n}{dx^n} x^m = \frac{\Gamma(m+1)}{\Gamma(m-n+1)} x^{m-n}. \tag{6.4}$$

Vanishing of the derivative for $m < n$ now follows automatically, owing to the divergence of the Gamma function for whole negative values of the argument. Equation (6.4), however, gives a definition of an object that exists for any values of parameters n and m, not necessarily whole numbers, so that

$$\frac{d^\nu}{dx^\nu} x^\mu = \frac{\Gamma(\mu+1)}{\Gamma(\mu-\nu+1)} x^{\mu-\nu}. \tag{6.5}$$

The fractional derivative of unity is given by

$$\frac{d^\nu}{dx^\nu} 1 = \frac{1}{\Gamma(1-\nu)} x^{-\nu}$$

and vanishes only for natural values of ν.

An interesting consequence of this rule of differentiation of powers is that the fractional derivatives of a constant do not vanish:

$$\frac{d^\nu}{dx^\nu} 1 = \frac{1}{\Gamma(1-\nu)} x^{-\nu}. \tag{6.6}$$

Exercise 6.2 Prove that Eq.(6.5) is exactly that following from the definition for the Riemann–Liouville derivative ${}_0D_x^\nu$ given by Eqs.(6.2) and (6.3) (note that here $a = 0$).

Hint: Use the definition of the Beta-function, $\mathrm{B}(p,q) = \frac{\Gamma(p)\Gamma(q)}{\Gamma(p+q)} = \int_0^1 u^{p-1}\,(1-u)^{q-1}du$.

A convenient property of the Riemann–Liouville derivative is connected to the fact that the Laplace representation of the fractional integral ${}_0I_t^\alpha$, being the convolution operator, is given by

$$\hat{L}\left\{{}_0I_t^\alpha f(t)\right\} = s^{-\alpha}\hat{L}\left\{f(t)\right\} \equiv s^{-\alpha} f(s), \tag{6.7}$$

where $\hat{L}$ denotes the operator of the Laplace transform, which for integer α corresponds to a known representation of repeated integrals.

Exercise 6.3 Use Eq.(6.7) to obtain the fractional integral of a power function x^μ.

The operators discussed above, i.e., the fractional derivatives ${}_0D_x^\nu$ with $0 < \nu < 1$, therefore have the Laplace representation $\hat{L}\left\{{}_0D_t^\nu f(t)\right\} = s^\nu \hat{L}\{f(t)\} \equiv s^\nu f(s)$ similar to that of the normal derivative $\hat{L}\left\{\frac{d}{dt} f(t)\right\} = s\,\tilde{f}(s) - f(0)$ but lacking the initial condition term, which is zero. Here we follow our convention to denote the Laplace-transformed function by a tilde provided that it is encountered in the same equation with the non-transformed initial condition.

Exercise 6.4 Use Eq.(6.7) to obtain the Laplace representation of the fractional derivative ${}_0D_t^\nu$ of arbitrary order ν. Note the dependence of the Laplace representation of derivatives on initial conditions.

While the fractional derivatives of power functions follow the familiar pattern for the "normal," integer derivatives, the result for an exponential may seem somewhat disappointing:

$$ {}_0D_x^\alpha e^x = e^x \frac{\gamma(-\alpha, x)}{\Gamma(-\alpha)}, $$

where $\gamma(-\alpha, x)$ is the incomplete Gamma function. However, we have to keep in mind that, compared to normal derivatives, the fractional ones contain an additional parameter a, the origin of integration, which in our previous examples was set to zero. If we let it tend to $-\infty$, the well-known form of the derivative of the exponential is restored:

$$ {}_{-\infty}D_x^\alpha e^x = e^x. $$

This special case of the Riemann–Liouville derivative ${}_aD_x^\alpha$ with $a \to -\infty$ is called the Weyl derivative. Just as the Riemann–Liouville derivative with $a = 0$ retains the known properties of "normal" derivatives under Laplace transform, the Weyl derivative has the familiar properties under Fourier transform

$$ \hat{F}\left\{ {}_{-\infty}D_x^\alpha f(x) \right\} = (ik)^\alpha \hat{F}\left\{ f(x) \right\} \equiv (ik)^\alpha f(k). $$

We return to this property in Chapter 7, on Lévy flights.

For more information about the mathematical side of the problem, consult Refs. [1–3].

6.2 Grünwald–Letnikov representation

The normal derivatives appear as limits of difference constructions, for example $\frac{df}{dx} = \lim_{\Delta x \to 0} \frac{f(x+\Delta x) - f(x)}{\Delta x}$. The integrals are also limits of the corresponding Riemann integral sums. The integro-differential operator of the fractional derivative has a representation in the form of an infinite sum, called the Grünwald–Letnikov formula. Thus, for fractional integration we have

$$ {}_aI_x^\alpha f(x) = {}_aD_x^{-\alpha} f(x) = \lim_{N \to \infty} \left[\left(\frac{x-a}{N} \right)^\alpha \sum_{j=0}^{N} \frac{\Gamma(j+\alpha)}{\Gamma(\alpha) j!} f\left(x - j\left(\frac{x-a}{N} \right) \right) \right]. $$

To understand how this works, take $\alpha = 1$, in which case the equation reproduces to the simple integral sum for $\int_a^x f(x')dx'$. For a general fractional derivative we get

$$ {}_aD_x^\alpha f(x) = \lim_{N \to \infty} \left[\left(\frac{N}{x-a} \right)^\alpha \sum_{j=0}^{N} (-1)^j \frac{\Gamma(\alpha+1)}{\Gamma(\alpha - j + 1) j!} f\left(x - j\left(\frac{x-a}{N} \right) \right) \right]. $$

Note that it is only for positive integer α that the difference representation of the derivative contains a fixed number of summands different from the number N of the points (separated by $\Delta x = (x-a)/N$) at which the function $f(x)$ is evaluated. This is because of the divergence of the Γ-function in the denominator. This is also the only case in which the explicit dependence on a vanishes. This formula gives a basis for several effective numerical algorithms for solutions of fractional equations.

6.3 Fractional diffusion equation

In the case of the power-law waiting-time distribution, our generalized Fokker–Planck equation, discussed at the beginning of this chapter, takes the form

$$\frac{\partial p(x,t)}{\partial t} = {}_0D_x^{1-\alpha}\left[-\mu_\alpha \frac{\partial}{\partial x} f(x)p(x,t) + K_\alpha \frac{\partial^2}{\partial x^2} p(x,t)\right]. \tag{6.8}$$

We consider here the one-dimensional situation with μ_a being the generalized mobility and K_a being the generalized (sub)diffusion coefficient (see Chapter 3).[1]

Let us now discuss two simple examples, one being the subdiffusive motion of the particle in one dimension under constant force $f = const$ (including the special case of free motion corresponding to $f = 0$), and the second example corresponding to the motion under linear restoring force, $f(x) = -kx$ (the generalized Ornstein–Uhlenbeck process).

Let us start from the case of the constant force

$$\frac{\partial p(x,t)}{\partial t} = {}_0D_x^{1-\alpha}\left[-\mu_\alpha f \frac{\partial}{\partial x} p(x,t) + K_\alpha \frac{\partial^2}{\partial x^2} p(x,t)\right]. \tag{6.9}$$

and discuss the behavior of the moments of displacement, $m_n(t) = \langle x^n(t)\rangle = \int_{-\infty}^{\infty} x^n\, p(x,t)\, dx$. The equations for the moments can be obtained by the multiplication of both parts of Eq.(6.9) by x^n and integration over x. The terms on the right-hand side of the equation are then rewritten using integration by parts, leading to

$$\int_{-\infty}^{\infty} x^n \frac{\partial}{\partial x} P(x,t)dx = -nm_{n-1}(t)$$

for $n \geq 1$ for the first term, and the repeated partial integration giving

$$\int_{-\infty}^{\infty} x^n \frac{\partial^2}{\partial x^2} P(x,t)dx = n(n-1)m_{n-2}(t)$$

for $n \geq 2$ for the second term of the equation. The zeroth moment of the distribution is, evidently, 1. The general equations for the moments of the distribution are thus given by

[1]Note that the memory kernel $M(t)$ in Eq.(5.25), corresponding for $t > 0$ to the rate of steps, evidently has the dimension of the inverse time for any waiting-time distribution. For the power-law waiting-time PDFs $\psi(t)$ given by Eq.(3.19) we get $M(t) \simeq \frac{1}{\Gamma(\alpha)} \frac{t^{\alpha-1}}{\tau^\alpha}$. On the other hand, the fractional integral does not contain the additional $\tau^{-\alpha}$ contribution and has a dimension $[\mathrm{T}^{1-\alpha}]$. Thus, the generalized mobility μ_α and the generalized diffusion coefficient K_α have to contain an additional $\tau^{-\alpha}$ factor compared to μ and D in Eq.(5.29).

$$\frac{d}{dt}m_n(t) = n\mu_\alpha f\, {}_0D_t^{1-\alpha}\, m_{n-1}(t) + n(n-1)\, K_\alpha\, {}_0D_t^{1-\alpha} m_{n-2}(t) \tag{6.10}$$

for $n \geq 2$, and

$$\frac{d}{dt}m_1(t) = \mu_\alpha f\, {}_0D_t^{1-\alpha}\, 1 \tag{6.11}$$

for the first moment.

In the case where $f = 0$, the first moment is zero, and the second one is given by the solution to the fractional differential equation

$$\frac{d}{dt}m_2(t) = 2K_\alpha\, {}_0D_t^{1-\alpha}\, 1. \tag{6.12}$$

Since the expression of the fractional derivative of the constant is known, we have

$$\frac{d}{dt}m_2(t) = 2K_\alpha \frac{1}{\Gamma(\alpha)} t^{\alpha-1},$$

which gives

$$m_2(t) = m_2(0) + 2K_\alpha \frac{1}{\alpha\Gamma(\alpha)} t^\alpha = m_2(0) + \frac{2K_\alpha}{\Gamma(\alpha+1)} t^\alpha.$$

If we consider the situation in which a particle starts at the origin at $t = 0$, the first term vanishes and

$$\left\langle x^2(t) \right\rangle_{f=0} = \frac{2K_\alpha}{\Gamma(\alpha+1)} t^\alpha, \tag{6.13}$$

where we have returned to the usual notation for a mean, and have denoted explicitly that the result is pertinent to the case $f = 0$. Let us now consider the case of a constant, non-vanishing force. In this case the first moment of the distribution does not vanish and is governed by Eq.(6.11), which differs from Eq.(6.12) only by the prefactor on the right-hand side. For a particle starting at the origin at $t = 0$ its solution thus reads

$$\langle x(t) \rangle_f = \frac{\mu_\alpha f}{\Gamma(\alpha+1)} t^\alpha. \tag{6.14}$$

Comparing Eq.(6.10) and Eq.(6.14) we see that

$$\langle x(t) \rangle_f = \frac{1}{2} \frac{\mu_\alpha}{K_\alpha} f \left\langle x^2(t) \right\rangle_{f=0}.$$

We note that in the cases where the system possesses true thermodynamic equilibrium, K_α and μ_α are connected by the generalized Einstein relation $K_\alpha/\mu_\alpha = k_B T$, so that

$$\langle x(t) \rangle_f = \frac{1}{2k_B T} f \left\langle x^2(t) \right\rangle_{f=0}. \tag{6.15}$$

Let us now consider the case of the parabolic potential (linear restoring force) described by the equation

$$\frac{\partial p(x,t)}{\partial t} = {}_0 D_t^{1-\alpha} \left[\mu_\alpha k \frac{\partial}{\partial x} (x p(x,t)) + K_\alpha \frac{\partial^2}{\partial x^2} p(x,t) \right]. \tag{6.16}$$

The equation for the first moment follows from multiplication of both parts of Eq.(6.16) by x and integration. Performing partial integration on the right-hand side gives

$$\frac{d}{dt} m_1(t) = -\tau^{-\alpha} \, {}_0 D_t^{1-\alpha} \, m_1(t), \tag{6.17}$$

where we have introduced the characteristic time τ defined by $\tau^{-\alpha} = k\mu_\alpha$. The solution to this equation is given through the Mittag-Leffler function, a special case of which we have already encountered in Chapter 3. Since Eq.(6.17) essentially has the form of a convolution, it can be solved by the Laplace transform, which reduces Eq.(6.17) to an algebraic equation

$$s \, \tilde{m}_1(s) - m_1(0) = -\tau^{-\alpha} \, s^{1-\alpha} \, \tilde{m}_1(s)$$

having the solution

$$\tilde{m}_1(s) = m_1(0) \frac{1}{s + \tau^{-\alpha} s^{1-\alpha}}.$$

Our next step is to find the inverse Laplace transform of the function

$$f(s) = \frac{1}{s} \frac{1}{1 + \tau^{-\alpha} s^{-\alpha}} = \frac{1}{s} \left(1 - \tau^{-\alpha} s^{-\alpha} + \tau^{-2\alpha} s^{-2\alpha} - \ldots \right) = \sum_{k=0}^{\infty} (-1)^k \tau^{-k\alpha} s^{-k\alpha - 1},$$

where we have simply expanded the fraction in the geometric series in $(\tau \, s)^{-\alpha}$. Now we can perform the term-per-term inverse Laplace transform using the well-known formula for the inverse Laplace transform of the power function $\hat{L}^{-1} \{ s^{-x} \} = t^{x-1} / \Gamma(x)$ to get

$$E_\alpha(z) = \sum_{k=0}^{\infty} \frac{z^k}{\Gamma(k\alpha + 1)}$$

is the Mittag-Leffler function. For $\alpha = 1$ it reduces to $E_1(z) = e^z$.

$$f(t) = \hat{L}^{-1} \{ f(s) \} = \sum_{k=0}^{\infty} \frac{(-1)^k \tau^{-k\alpha} t^{k\alpha}}{\Gamma(k\alpha + 1)}.$$

The series $E_\alpha(z) = \sum_{k=0}^{\infty} \frac{z^k}{\Gamma(k\alpha+1)}$ defines the Mittag-Leffler function, so that $f(t) = E_\alpha \lfloor -(t/\tau)^\alpha \rfloor$, and the solution to Eq.(6.17) reads

$$m_1(t) = m_1(0)E_\alpha\left[-\left(t/\tau\right)^\alpha\right].$$

The solution to the fractional differential equation

$$\frac{dy}{dx} = {}_0D_x^{1-\alpha}y$$

reads

$$y(x) = y_0 E_\alpha\left[-x^\alpha\right]$$

The behavior of the Mittag-Leffler functions $E_\alpha(-t^\alpha)$ is shown in Fig. 6.1. The function $E_1(-t) = \exp(-t)$ is a decaying exponential. Those with $0 < \alpha \leq 1$ behave as stretched exponentials for $t << 1$, so that $E_\alpha(-t^\alpha) \sim \exp\left(-\frac{t^\alpha}{\Gamma(1+\alpha)}\right)$, but cross over to a power-law behavior

$$E_\alpha(-t^\alpha) \sim \frac{1}{\Gamma(1-\alpha)t^\alpha} \tag{6.18}$$

for large absolute values of the argument. This is elucidated by a double logarithmic plot of $E_{1/2}(-t^{1/2})$ in Fig. 6.2. This function can be expressed through the error function: $E_{1/2}(-t^{1/2}) = e^t\ \mathrm{erfc}(\sqrt{t})$.

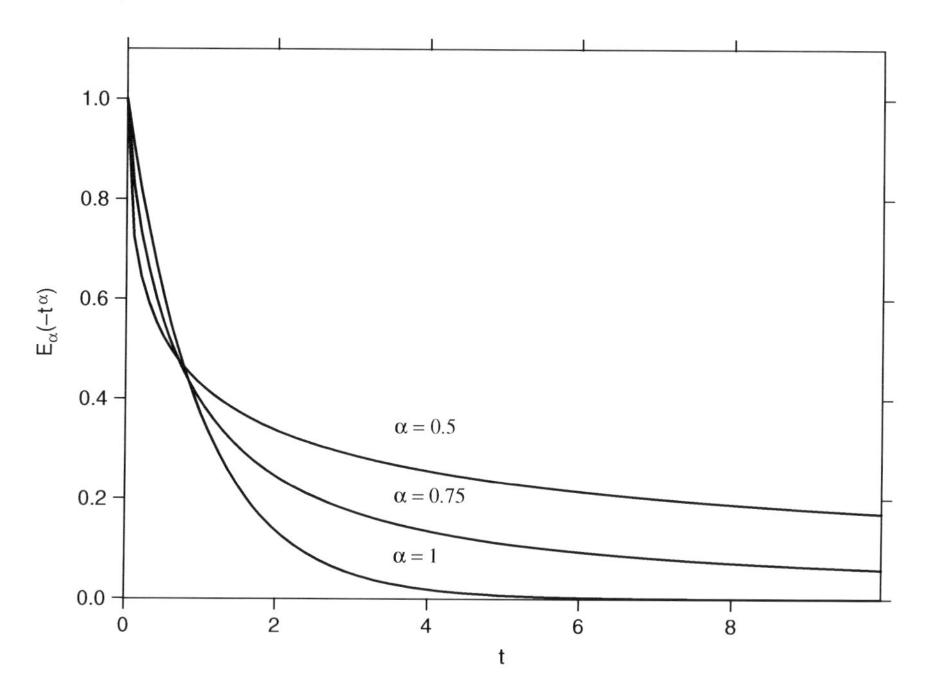

Fig. 6.1 *Mittag-Leffler functions $E_\alpha(-t^\alpha)$ for $\alpha = 0.5$, 0.75, and 1. Note that, with smaller index α, the slower is the overall decay.*

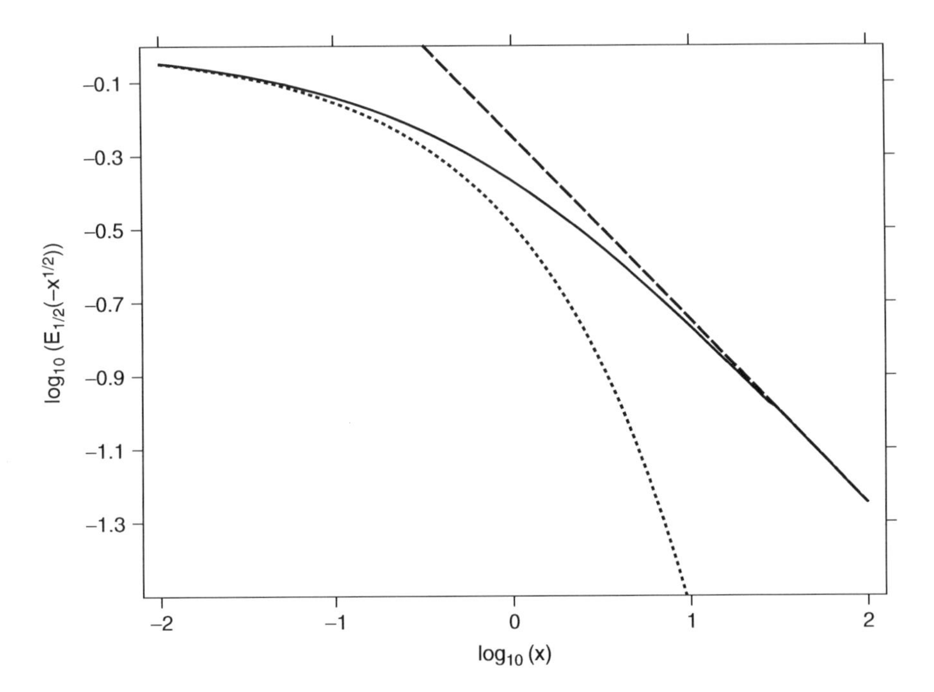

Fig. 6.2 *The Mittag-Leffler function $E_{1/2}(-t^{1/2})$ shown on double logarithmic scales together with its small- and large-t asymptotic behavior (stretched exponential: dotted line; power-law: dashed line).*

> **Exercise 6.5** Use Eq.(6.16) to obtain the closed equation for the temporal evolution of the second moment $m_2(t) = \langle x^2(t) \rangle$ of the distribution. Solve this equation to show that $\langle x^2(t) \rangle$ behaves as $\langle x^2(t) \rangle = x_{eq}^2 + \left(x_0^2 - x_{eq}^2\right) E_\alpha[-2\,(t/\tau)^\alpha]$ and find the corresponding parameters x_0^2 and x_{eq}^2.

A note on time-dependent forces should be made here. In the continuous limit the master equation (5.18) derived under the assumption that the external force affects only the transition probabilities, but not waiting times, reduces to the following fractional Fokker–Planck equation in the form

$$\frac{\partial p(x,t)}{\partial t} = \left[-\mu_\alpha \frac{\partial}{\partial x} f(x,t) + K_\alpha \frac{\partial^2}{\partial x^2}\right] {}_0D_x^{1-\alpha} p(x,t) \tag{6.19}$$

with the time-dependent Fokker–Planck operator *preceding* the fractional derivative. The sequence of the operators did not matter in Eq.(6.9) since they acted on different variables of $p(x,t)$ and therefore commuted. Now this is no longer the case and the order of the operators is fixed. Equation (6.19) allows us to obtain moments of the corresponding distributions. For example, the evolution of the first moment is given by

$$\frac{d}{dt} m_1(t) = \mu_\alpha f(t)\, {}_0D_t^{1-\alpha}\, 1$$

(cf. Eq.(6.11)), which reproduces the results of Sec. 4.5 (see Ref.[4] for a discussion of the behavior of the second moment).

6.4 Eigenfunction expansion

Fractional Fokker–Planck equations with time-independent forces allow for eigenfunction expansions just like the normal ones do. One of the reasons for using this approach is that it is absolutely parallel to the one used in solving Schrödinger equations in quantum mechanics. Here again both effective approximations and effective numerical implementations exist.

Just as in the case of a time-dependent Schrödinger equation, we look for the solution in the form $p(x,t) = \sum_{n=0}^{\infty} a_n(x_0,t_0) p_n(x,t)$ with expansion coefficients $a_n(x_0,t_0)$ depending on the initial conditions and use the method of separation of variables:

$$p_n(x,t) = \phi_n(x) T_n(t)$$

for a given eigenmode n. We get the decoupled set for the equation for the temporal and spatial eigenfunctions

$$\hat{L}_{FP} \phi_n(x) = -\lambda_n \phi_n(x)$$

and

$$\frac{d}{dt} T_n(t) = -{}_0D_t^{1-\alpha} \lambda_n T_n(t),$$

where the first equation is that for the eigenfunctions of the usual Fokker–Planck operator, and the second equation has a solution that we have already met above:

$$T_n(t) = E_\alpha(-\lambda_n t^\alpha)$$

Example 6.1 The simplest example of eigenfunction expansion is the diffusion problem on the interval with absorbing boundaries, described by the fractional diffusion equation

$$\frac{\partial p(x,t)}{\partial t} = K_{\alpha_0} D_x^{1-\alpha} \frac{\partial^2}{\partial x^2} p(x,t) \tag{6.20}$$

with $p(-L/2,t) = p(L/2,t) = 0$. In this case the Fokker–Planck operator $\hat{L}_{FP}$ is simply Laplacian, and its eigenfunctions vanishing at the boundaries are known to us from the solution to the problem of oscillating string in mechanics, or from the problem of a particle in a box in quantum mechanics. The solutions of $\Delta\phi_n(x) = -\lambda_n \phi_n(x)$ that vanish at $x = \pm L/2$ are $\phi_n = \cos(\pi(2n+1)x/L)$.

Exercise 6.6 Use eigenfunction decomposition to obtain the time evolution of the probability $p(x,t)$ of finding a particle at position x on an interval $[-L/2, L/2]$ with absorbing boundaries, given by the solution to the fractional diffusion equation (6.20). The initial position of the particle is equally distributed within the interval, $p(x,0) = const = \frac{1}{L}$.

Show that

$$p(x,t) = \sum_{m=1}^{\infty} (-1)^m \frac{4}{L\pi(2m+1)} \cos\left(\frac{\pi(2m+1)x}{L}\right) E_\alpha\left(-\frac{\pi^2(2m+1)^2}{L^2} K_\alpha t^\alpha\right).$$

Find the survival probability of the particle on the interval $\Phi(t) = \int_{-L/2}^{L/2} p(x,t)\,dx$.

This result will be useful in Chapter 9.

In the case of normal diffusion ($E_1(-x^1) = e^{-x}$), the expressions for $p(x,t)$ and for $\Phi(t)$ reduce to

$$p(x,t) = \sum_{m=1}^{\infty} (-1)^m \frac{4}{L\pi(2m+1)} \cos\left(\frac{2\pi(2m+1)x}{L}\right) \exp\left(-\frac{\pi^2(2m+1)^2}{L^2} Dt\right) \tag{6.21}$$

and

$$\Phi(t) = \sum_{m=1}^{\infty} \frac{8}{\pi^2(2m+1)^2} \exp\left(-\frac{\pi^2(2m+1)^2}{L^2} Dt\right). \tag{6.22}$$

Let us now turn to more complex situations, where the force term in the Fokker–Planck operator is present. The situation is especially advantageous in one-dimensional problems where a simple variable transformation changes the non-Hermitian Fokker–Planck operator to a Hermitian Schrödinger one. To make the transformation we first put the Fokker–Planck operator into a dimensionless form by the corresponding choice of variables (i.e., by changing to the reduced coordinate x and to reduced time t), so that

$$\frac{\partial}{\partial x}\left(\frac{\partial U}{\partial x} f_n(x)\right) + \frac{\partial^2}{\partial x^2} f_n(x) = -\lambda_n f_n(x)$$

and make a change of variable according to $\psi(x) = e^{-U(x)/2} f(x)$. The equation for ψ then admits variable separation. Its spatial part fulfills:

$$\hat{H}\psi_n(x) = \lambda_n \psi_n(x)$$

with

$$\hat{H}\psi(x) = -\frac{\partial^2}{\partial x^2}\psi(x) + V(x)\psi(x)$$

and with the effective potential[2]

$$V(x)=\frac{1}{4}\left(\frac{\partial U}{\partial x}\right)^2+\frac{1}{2}\frac{\partial^2 U}{\partial x^2}.$$

Just as in quantum mechanics, the solution (under the corresponding choice of units) therefore reads:

$$p(x,t|x_0,t_0)=e^{U(x_0)/2-U(x)/2}\sum_{n=0}^{\infty}\psi_n(x)\psi_n(x_0)E_\alpha(-\lambda_n t^\alpha).$$

The situation is especially simple for the particle in the harmonic potential, where the corresponding effective potential in the Schrödinger equation is again harmonic. The reasonable choice of the reduced coordinates here is: $t \to t/\tau$, $x \to x\sqrt{k/(k_BT)}$. The Fokker–Planck operator then corresponds to the potential $U(x) = x^2/2$, and the Schrödinger potential is $V(x)=\frac{1}{4}x^2+\frac{1}{2}$. The eigenvalues of the corresponding Schrödinger equation for a harmonic oscillator with the formal parameters $\hbar = 1$, $m = 1/2$, and $\omega = 1$ are then $\lambda_n = E_n - \frac{1}{2} = \hbar n = n$ (the fact that $\lambda_n = E_n - \frac{1}{2}$ is due to the additive correction to the potential), and the corresponding eigenfunctions are given by

$$\psi_n(x)=\left(\frac{1}{\sqrt{2\pi}n!2^n}\right)^{1/2}H_n\left(\frac{x}{\sqrt{2}}\right)e^{-x^2/4}$$

where $H_n(z)$ are the Hermite polynomials.

The final solution then reads

$$p(x,t)=\sum_{n=0}^{\infty}\left(\frac{1}{\sqrt{2\pi}n!2^n}\right)H_n\left(\frac{x}{\sqrt{2}}\right)H_n\left(\frac{x_0}{\sqrt{2}}\right)e^{-x^2/2}E_\alpha(-nt^\alpha).$$

For $t\to\infty$ this solution is dominated by the term with $n = 0$ and converges to

$$p(x,t)=\frac{1}{\sqrt{2\pi}}e^{-x^2/2},$$

as the representation of the Boltzmann distribution, which in natural units reads

$$p(x,t)=\frac{1}{\sqrt{2\pi k_BT/k}}\exp\left(-\frac{kx^2}{2k_BT}\right).$$

[2]For each U this potential has the form $V(x) = W'(x) + W^2(x)$ so that the corresponding Hamiltonian of the Schrödinger equation can be written as $H = -\left(\frac{d}{dx}+W\right)\left(\frac{d}{dx}-W\right)$, corresponding to so-called supersymmetric quantum mechanics; see, e.g., Ref.[5] for a basic review. Such Schrödinger equations always possess a zero eigenvalue provided that the "superpotential" W does not decay in infinity too fast, i.e., for $\int_0^x W(x')dx' \to \infty \int_0^x W(x')dx' \to -\infty$ for $x\to\pm\infty$. This property, as we see at the end of this section, corresponds to the fact that the system possesses a thermodynamic equilibrium state.

6.5 Subordination and the forms of the PDFs

We now turn to the full solution to the fractional diffusion equations for some important special cases. Of course, just as for the master equations discussed in the previous chapters, such solutions can be easily obtained by using the integral representation (Eqs.(5.20) and (5.21)). For the case of the power-law memory kernel corresponding to $M(s) \propto s^{-\alpha}$ the function $T(\omega, t)$ is given by its Laplace transform in its second variable, which reads $T(\omega, s) = s^{\alpha-1} \exp[-\omega s^{\alpha}]$ (we here use ω for the first temporal variable since the symbol τ is used in this chapter to denote a different quantity). The explicit form of $T(\omega, t)$ can be obtained, e.g., for $\alpha = 1/2$, for which the inverse Laplace transform in s is easily performed and gives $T(\omega, t) = \exp\left(-\omega^2/4t\right)/\sqrt{\pi t}$, a "half-Gaussian," which allows us to easily transform the results. The solution for the PDF is given by the integral formula of subordination. This explicit form is used in our graphical examples (Figs. 6.1 and 6.2).

Let us consider, e.g., the free diffusion case, i.e., that without external force. In this case

$$f(x, \omega) = \frac{1}{\sqrt{4\pi K\omega}} \exp\left(-\frac{x^2}{4K\omega}\right),$$

and therefore $p(x, t)$ is given by the integral

$$p(x,t) = \int_0^\infty \frac{1}{\sqrt{4\pi K\omega}} \exp\left(-\frac{x^2}{4K\omega}\right) \frac{1}{\sqrt{\pi t}} \exp\left(-\frac{\omega^2}{4t}\right) d\omega, \tag{6.23}$$

which can be easily evaluated numerically. This is exactly what was done when plotting Fig. 3.5 in Chapter 3. Here we repeat the corresponding plot in Fig. 6.3. The PDF for the fractional diffusion is shown for $K = 1$ at time $t = 1$.

The same approach is used to obtain the pictures for the fractional Ornstein–Uhlenbeck process, for which

$$f(x,t) = \frac{1}{\sqrt{2\pi K\tau\left[1 - e^{-2t/\tau}\right]}} \exp\left(-\frac{\left(x - x_0 e^{-t/\tau}\right)^2}{2K\tau\left[1 - e^{-2t/\tau}\right]}\right), \tag{6.24}$$

with x_0 being the initial condition at $t = 0$ (see Fig. 6.4).

Exercise 6.7 Check that Eq.(6.24) is indeed the solution to the equation $\frac{\partial f(x,t)}{\partial t} = \mu k \frac{\partial}{\partial x}(x f(x,t)) + K \frac{\partial^2}{\partial x^2} f(x,t)$, and express the relaxation time τ through the parameters of the problem.

Another important use of the subordination transformation is discussed in Example 6.2.

Example 6.2 Let us reconsider the problem discussed in Example 6.1 and find the survival probability $P_L(t) = \int_{-L/2}^{L/2} p(x,t)dx$ of a particle on an interval $[-L/2, L/2]$

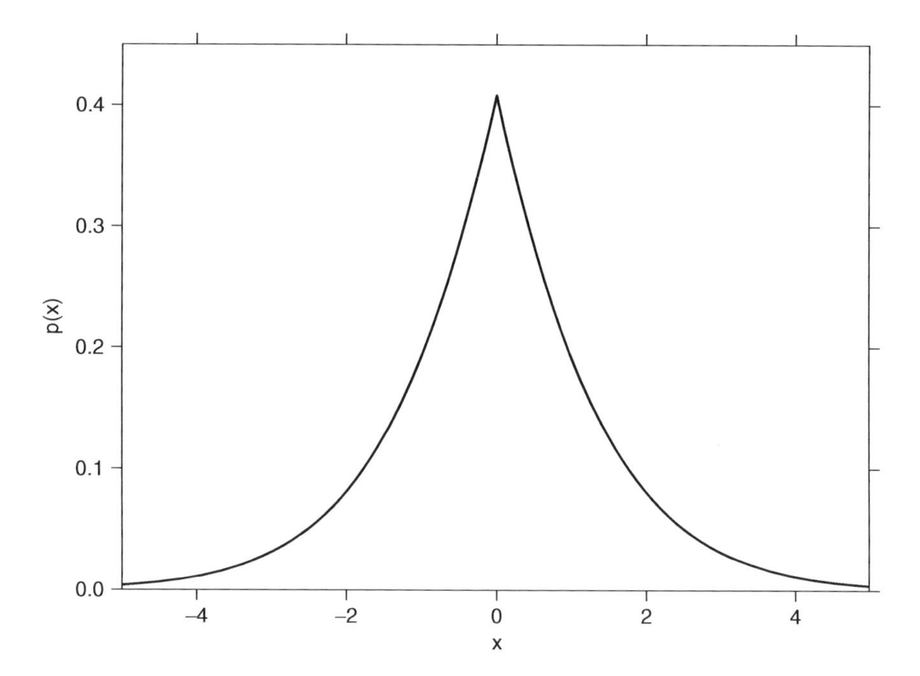

Fig. 6.3 *The PDF of the displacements of a walker in a CTRW with* $\psi(t) \propto t^{-3/2}$. *Note the cusp typical of all subdiffusive CTRWs.*

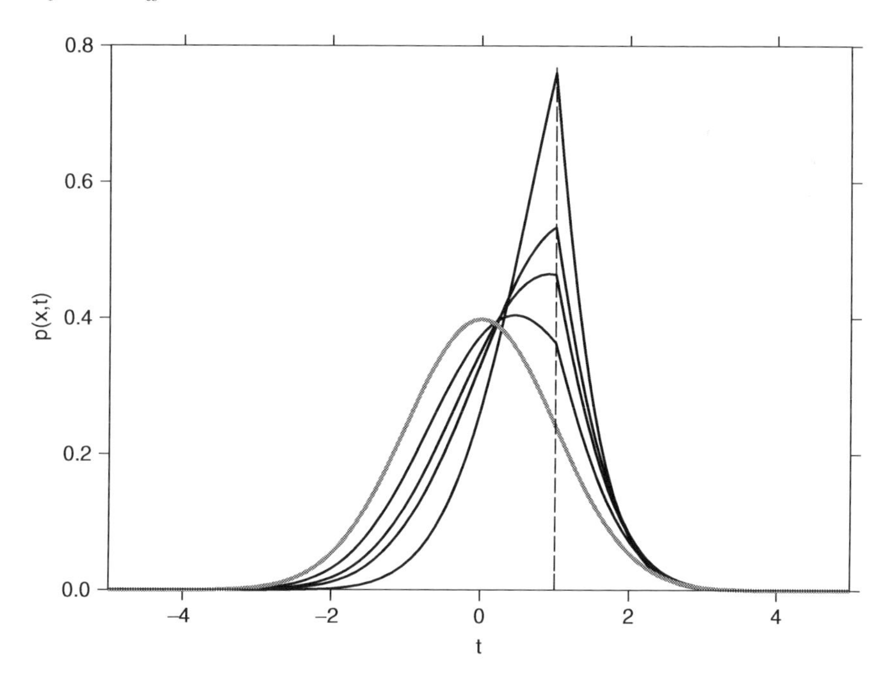

Fig. 6.4 *PDFs of the particle positions for the fractional Ornstein–Uhlenbeck process with* $\alpha = 1/2$, $\tau_0 = 1$. *The initial position of the particle is* $x_0 = 1$. *The black curves correspond to the times* $t = 0.1$, 0.5, 1, *and* 2, *respectively (the longer the time, the lower the maximum of the curve). Note the cusp at the initial position of the particle. The gray curve represents the final Gaussian distribution.*

with absorbing boundaries. Now we will apply the method of subordination. Here it is convenient to make all calculations in the Laplace domain, and to return to the time domain only at the very last step of our calculation. The results of this discussion will be crucial in Chapter 9.

The probability density of particle positions is given by the solution to the fractional diffusion equation (6.20)

$$\frac{\partial p(x,t)}{\partial t} = K_\alpha {}_0D_x^{1-\alpha} \frac{\partial^2}{\partial x^2} p(x,t)$$

with boundary conditions $p(-L/2,t) = p(L/2,t) = 0$. The initial position of a particle is equally distributed within the interval, $p(x,0) = const = \frac{1}{L}$.

We first have to solve the customary analog of Eq.(6.20),

$$\frac{\partial f(x,t)}{\partial t} = K \frac{\partial^2}{\partial x^2} f(x,t) \tag{6.25}$$

with the same initial and boundary conditions. Applying the Laplace transform to both sides of Eq.(6.25), we get an ordinary linear differential equation

$$sf(x,s) - \frac{1}{L} = \frac{d^2}{dx^2} Kf(x,s), \tag{6.26}$$

which can be easily integrated. As usual, this is done by assuming the solution to be a sum of a general solution to the homogeneous equation, $A\exp\left(\sqrt{\frac{s}{K}}x\right) + B\exp\left(-\sqrt{\frac{s}{K}}x\right)$, and of a special solution to an inhomogeneous equation, e.g., $1/sL$. The solution to Eq.(6.26) fulfilling boundary conditions $f(-L/2,s) = f(L/2,s) = 0$ reads:

$$f(x,s) = \frac{1}{Ls}\left[1 - \frac{\cosh\left(\sqrt{\frac{s}{K}}x\right)}{\cosh\left(\sqrt{\frac{s}{K}}\frac{L}{2}\right)}\right].$$

Integrating this solution over the interval $[-L/2, L/2]$, we get

$$\tilde{\Phi}(s) = \int_{-L/2}^{L/2} f(x,s)dx = \frac{1}{s} - \frac{2\sqrt{K}}{Ls^{3/2}} \tanh\left(\sqrt{\frac{s}{K}}\frac{L}{2}\right). \tag{6.27}$$

Equation (6.22) now follows from Eq.(6.27) when applying the known formula

$$\tanh\frac{\pi x}{2} = \frac{4x}{\pi}\sum_{k=0}^{\infty}\frac{1}{(2k+1)^2 + x^2}$$

(see Eq.1.421.2 of Ref.[6]); using this, we obtain

$$\tilde{\Phi}(s) = \frac{1}{s} - \frac{2\sqrt{K}}{Ls^{3/2}} \tanh\left(\sqrt{\frac{s}{K}}\frac{L}{2}\right) = \frac{1}{s} - \frac{1}{s}\frac{8}{\pi^2}\sum_{m=0}^{\infty}\frac{1}{(2m+1)^2 + \frac{s}{K}\frac{L^2}{\pi^2}}. \tag{6.28}$$

Exercise 6.8 Prove that the inverse Laplace transform of Eq.(6.28) leads to Eq.(6.22).

Hint: Each term of the series in Eq.(6.28) corresponds to an exponential, and the $1/s$ preceding it denotes that the integral of it is taken over time. Note that $\sum_{m=0}^{\infty} \frac{1}{(2m+1)^2} = \frac{\pi^2}{8}$.

The transition to the fractional domain can be performed on the level of Eq.(6.27) or on the level of Eq.(6.28). The survival probability $\Phi(t)$ in the fractional case and that for customary diffusion, $\tilde{\Phi}(t)$, are connected in the Laplace domain via the analog of Eq.(5.24):

$$\Phi(s) = \frac{1}{sM(s)} \tilde{\Phi}\left(\frac{1}{M(s)}\right) = \frac{1}{s^{1-\alpha}} \tilde{\Phi}\left(s^{\alpha}\right)$$

Applying this to Eq.(6.28) and using the same trick as in Exercise 6.8, we obtain

$$\Phi(t) = \sum_{m=1}^{\infty} \frac{8}{\pi^2(2m+1)^2} E_{\alpha}\left(-\frac{\pi^2(2m+1)^2}{L^2} K t^{\alpha}\right). \tag{6.29}$$

We can also simply apply the integral formula of subordination to Eq.(6.22). Verification of this is left to the reader.

Exercise 6.9 Obtain the result for $\Phi(t)$, Exercise 6.6, by applying the integral formula of subordination to Eq.(6.22).

References

[1] K. Oldham and J. Spanier. *The Fractional Calculus*, New York: Academic Press, 1974

[2] K.S. Miller and B. Ross. *An Introduction to the Fractional Calculus and Fractional Differential Equations*, New York: John Wiley and Sons, Inc., 1993

[3] I. Podlubny. *Fractional Differential Equations*, New York: Academic Press, 1999

[4] I.M. Sokolov and J. Klafter. *Phys. Rev. Lett.* **97**, 140602 (2006).

[5] L. E. Gendenshtein and I.V. Krive, *Uspekhi Fiz. Nauk.* **146**, 553 (1985) (English version: *Sov. Phys. Usp.* **28** (8), 645 (1985))

[6] I.S. Gradstein and I.M. Ryzhik. *Table of Integrals, Series and Products*, Academic Press, Boston, 1994

Further reading

I.M. Sokolov, J. Klafter, and A. Blumen. "Fractional Kinetics," *Physics Today*, November 2002, p. 48

R. Metzler and J. Klafter. *Phys. Reports* **339**, 1 (2000)

H. Risken. *The Fokker-Planck Equation*, 2nd Edition, Berlin: Springer, 1996

B.J. West, M. Bologna, and P. Grigolini. *Physics of Fractal Operators*, New York: Springer, 2003

7
Lévy flights

"Either you'll find something solid to stand on or you'll be taught how to fly."

Richard Bach

Although some examples of random walks with diverging MSD per step were considered in Sec. 1.6, in the subsequent chapters we mostly dealt with random walks possessing such a moment. Now we return to such processes (the so-called Lévy flights) and discuss them in some detail.

As we have seen, the symmetric random walks with the step length distribution following a power law $p(x) \propto \frac{A}{|x|^{1+\alpha}}$ with $0 < \alpha \leq 2$ lead after $n >> 1$ steps to the distribution of displacements possessing the characteristic function of the form $f(k) = \exp(-a|k|^{\alpha})$, with a being some constant. In what follows, we consider more general cases of such distributions. The realizations of steps in such walks are shown in Fig. 7.1.

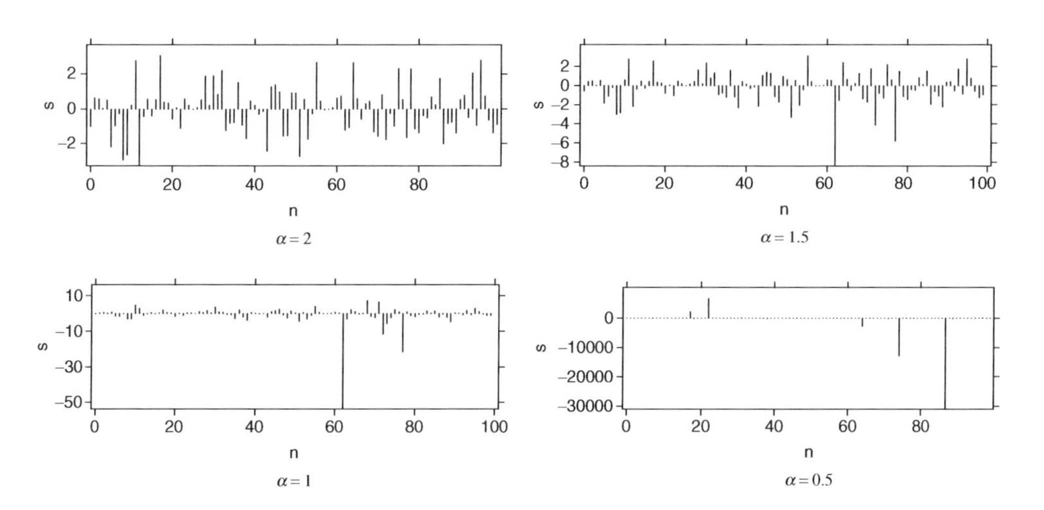

Fig. 7.1 *The realizations of the first 100 increments s_n of symmetric random flights distributed according to Lévy laws with $\alpha = 2$ (Gaussian), 1.5 (Holtsmark), 1.0 (Cauchy), and 0.5. Note the difference in scales. Note also that the smaller α is, the more the overall process is dominated by rare but large events.*

7.1 General Lévy distributions

Let us consider a more general situation of an asymmetric walk, characterized by the power-law fat tail of the PDF

$$p(x) \cong \begin{cases} \frac{A_+}{x^{1+\alpha}} & \text{for } x \to \infty \\ \frac{A_-}{(-x)^{1+\alpha}} & \text{for } x \to -\infty \end{cases}, \tag{7.1}$$

where its cumulative distribution function behaves so that

$$\begin{aligned} x^\alpha [1 - F(x)] &\to \frac{A_+}{\alpha} \quad \text{for } x \to \infty \\ (-x)^\alpha F(x) &\to \frac{A_-}{\alpha} \quad \text{for } x \to -\infty \end{aligned}. \tag{7.2}$$

A general form of the characteristic function to which a sum of many independent variables with power-law fat-tailed distributions may tend reads:

$$f_{\alpha,\beta,\mu,\sigma}(k) = \exp\left(-\sigma^\alpha |k|^\alpha (1 - i\beta\omega(k,\alpha)\,\mathrm{sign}(k)) + i\mu k\right) \tag{7.3}$$

with

$$\omega(k,\alpha) \cong \begin{cases} \tan\frac{\pi\alpha}{2} & \text{for } \alpha \neq 1 \\ -\frac{2}{\pi}\ln|k| & \text{for } \alpha = 1 \end{cases}.$$

Equation (7.3) corresponds to a four-parametric family of functions $L_{\alpha,\beta,\mu,\sigma}(x)$, where the parameters are the Lévy index $\alpha \in [0,2]$, the skewness (asymmetry) parameter $\beta \in [-1,1]$, the scale parameter (width of the distribution) $\sigma > 0$, and the shift parameter μ. The last two parameters can be removed from the list by considering the normalized variable $y = \frac{x-\mu}{\sigma}$, so that $L_{\alpha,\beta,\mu,\sigma}(x) = L_{\alpha,\beta,0,1}\left(\frac{x-\mu}{\sigma}\right)$. In what follows, the values of μ and σ are set to $\mu = 0$ and $\sigma = 1$, respectively, and the notation $L_{\alpha,\beta}(x)$ is used. The PDFs $L_{\alpha,\beta}(x)$ and $L_{\alpha,-\beta}(x)$ are symmetric with respect to the origin of the coordinates, $L_{\alpha,\beta}(x) = L_{\alpha,-\beta}(-x)$. The parameter β of the Lévy distribution can be expressed through the constants A_+ and A_- : $\beta = \frac{A_+ - A_-}{A_+ + A_-}$.

Exercise 7.1 Prove that the parameter β of the Lévy distribution is given by $\beta = \frac{A_+ - A_-}{A_+ + A_-}$, where A_+ and A_- are defined in Eq.(7.1).

Hint: Present the function $p(x)$ as a sum of an even and an odd part, $p(x) = \frac{p(x)+p(-x)}{2} + \frac{p(x)-p(-x)}{2}$. Then use the trick analogous to Eq.(1.13).

The characteristic function, Eq.(7.3), can be presented in a different way:

$$f(k) = \exp\left(-\frac{1}{\cos(\pi\alpha/2)}\,|k|^\alpha\, e^{-i\pi\gamma\,\mathrm{sign}(k)/2}\right)$$

(written here for $\alpha \neq 1$) with $\mathrm{sign}(k) = k/|k|$ being the sign of k, and the new parameter γ is introduced according to the condition

$$\tan\frac{\pi\gamma}{2} = \beta\tan\frac{\pi\alpha}{2} \tag{7.4}$$

(note that for $0 < \alpha < 1$ we have $0 \le |\gamma| \le \alpha$ and for $1 < \alpha < 2$ it follows that $0 \le |\gamma| \le 2 - \alpha$).[1]

The one-sided distributions with $0 < \alpha < 1$ and $\beta = 1$, those with characteristic function

$$f_{\alpha,\beta}(k) = \exp\left(-|k|^{\alpha}\left(1 - i\tan\frac{\pi\alpha}{2}\,\mathrm{sign}(k)\right)\right), \tag{7.5}$$

are concentrated on the positive half-line. Here the transition from the Fourier to the Laplace representation is convenient.

Exercise 7.2 Show that the Laplace transform of $L_{\alpha,1}(x)$ reads

$$\hat{L}\left\{L_{\alpha,1}(x)\right\} = \int_0^{\infty} e^{-sx} L_{\alpha,1}(x)\,dx = \exp\left(-\frac{s^{\alpha}}{\cos(\pi\alpha/2)}\right) \tag{7.6}$$

Hint: Perform integration in the complex plain!

The behavior of the one-sided Lévy distribution $L_{\alpha,1}(x)$ close to zero is therefore

$$L_{\alpha,1}(x) = C_1 x^{-1-\frac{1}{2}\frac{\alpha}{1-\alpha}} \exp\left(-\frac{C_2}{x^{\frac{\alpha}{1-\alpha}}}\right) \tag{7.7}$$

with $C_1 = \dfrac{\alpha^{\frac{1}{2(1-\alpha)}}\left(\cos\frac{\pi\alpha}{2}\right)^{-\frac{1}{2(1-\alpha)}}}{\sqrt{2\pi(1-\alpha)}}$ and $C_2 = (1-\alpha)\alpha^{\frac{\alpha}{1-\alpha}}\left(\cos\frac{\pi\alpha}{2}\right)^{-\frac{1}{1-\alpha}}$.

Exercise 7.3 Prove Eq.(7.7).
Hint: It is easier to go from Eq.(7.7) to Eq.(7.6) than the other way around. Note that $x \to 0$ corresponds to $s \to \infty$.

For the Lévy–Smirnov distribution ($\alpha = 1/2$), Eq.(7.7) gives the representation of the distribution in the whole domain of x.

[1]This is also the form preferred by Feller [1], who also omits the prefactor $\cos\frac{\pi\alpha}{2}$ and writes $f(k) = \exp\left(-|k|^{\alpha} e^{i\pi\gamma\,\mathrm{sign}(k)/2}\right)$. This form differs from the canonical one by its scaling parameter. Denoting the inverse Fourier transform of this function by $L(x;\alpha,\gamma)$ we see that $L(x;\alpha,\gamma) = L_{\alpha,\beta}\left(\frac{x}{\cos(\pi\alpha/2)}\right)$ with β given by Eq.(7.3). This notation is especially convenient to get series expansions:

$$L(x;\alpha,\gamma) = \frac{1}{\pi x}\sum_{k=1}^{\infty}\frac{\Gamma(k\alpha+1)}{k!}(-x^{-\alpha})^k \sin\left(\frac{\kappa\pi(\gamma-\alpha)}{2}\right) \quad \text{for } 0 < \alpha < 1$$

and

$$L(x;\alpha,\gamma) = \frac{1}{\pi x}\sum_{k=1}^{\infty}\frac{\Gamma(k/\alpha+1)}{k!}(-x)^k \sin\left(\frac{\kappa\pi(\gamma-\alpha)}{2\alpha}\right) \quad \text{for } 1 < \alpha < 2.$$

Both expansions are valid for $x > 0$; for $x < 0$ we should note that $L(-x;\alpha,\gamma) = L(x;\alpha,-\gamma)$.

Table 7.1 Some special Lévy distributions for rational values of α.

α	β	Name	Analytical expression
2	arbitrary	Gaussian	$\frac{1}{\sqrt{2\pi}}\exp\left(-\frac{x^2}{2}\right)$
3/2	0	Holtsmark	–
1	0	Cauchy	$\frac{1}{\pi}\frac{1}{1+x^2}$
2/3	0	–	$\sqrt{\frac{3}{\pi}}\frac{1}{6x}\exp\left(\frac{2}{27x^2}\right)W_{-\frac{1}{2},\frac{1}{6}}\left(\frac{4}{27x^2}\right)$
1/2	0	–	$\sqrt{\frac{2}{\pi x}}\left\{\cos\left(\frac{1}{4x}\right)\left[1-2C\left(\frac{1}{\sqrt{2\pi x}}\right)\right]+\sin\left(\frac{1}{4x}\right)\left[1-2S\left(\frac{1}{\sqrt{2\pi x}}\right)\right]\right\}$
1/2	1	Lévy–Smirnov	$\frac{1}{\sqrt{2\pi x^3}}\exp\left(-\frac{1}{2x}\right)$
1/3	1	–	$\frac{1}{\pi}\left(\frac{2}{3^{7/6}x}\right)K_{1/3}\left(\frac{2^{5/2}}{3^{9/4}\sqrt{x}}\right)$

In general, Lévy PDFs build a distinct family of special functions, whose properties are well known. As shown in Ref.[2], the PDFs of Lévy distributions for rational values of index α can be expressed in terms of generalized hypergeometric functions. Some of these distributions have special names (such as Gaussian, Cauchy, Holtsmark, or Lévy–Smirnov distributions), and some of them have close analytical forms in terms of elementary or "simpler" special functions (see Table 7.1). The PDFs of the Gaussian, Holtsmark, and Cauchy distributions as well as that of the symmetric distribution of index $\alpha = 1/2$ are shown in Fig. 7.2. Figure 7.3 gives the PDFs of some asymmetric distributions with $\alpha = 3/2$ and $\alpha = 1/2$.

The forms of the probability densities given correspond to the inverse Fourier transforms of characteristic functions given by $f(k) = \exp(-|k|^\alpha)$ for the symmetric distributions and to the inverse Laplace transforms of $f(u) = \exp\left(-\frac{u^\alpha}{\cos(\pi\alpha/2)}\right)$ for asymmetric ones. Here $K_\nu(x)$ is the modified Bessel function, $C(x) = \int_0^x \cos\left(\pi u^2/2\right)du$ and $S(x) = \int_0^x \sin\left(\pi u^2/2\right)du$ are the Fresnel integrals, and $W_{a,b}(x)$ is a Whittaker function.[2] A somewhat more complicated expression, involving Lommel functions, exists for a symmetric Lévy distribution with $\alpha = 1/3$ [2].

[2]The result for $L_{2/3,0}(x)$ was first derived by V.M. Zolotarev in 1954 and contained misprints. It was widely reproduced in the literature, but evidently never used, since the error was noticed and corrected only 48 years later [2].

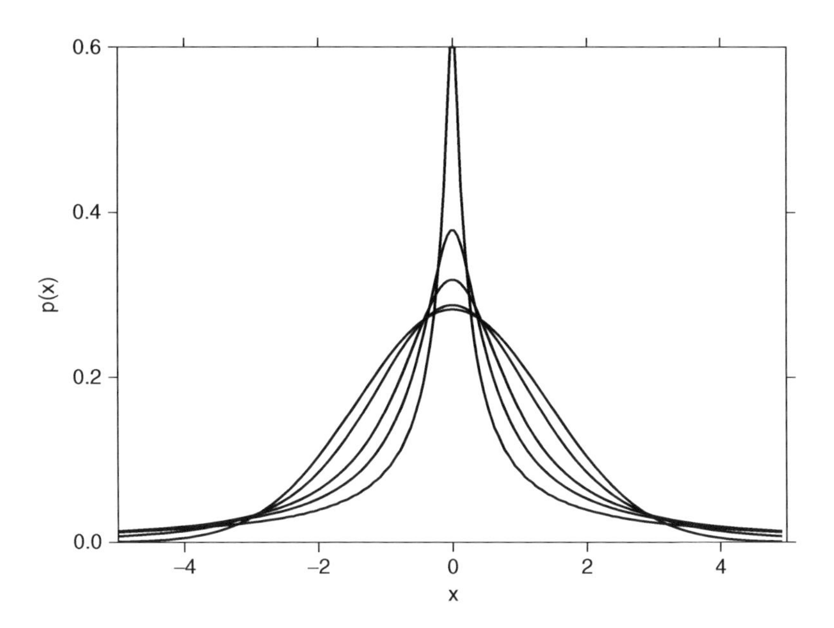

Fig. 7.2 *The PDFs of symmetric Lévy laws for* $\alpha = 0.5$, $\alpha = 0.75$, $\alpha = 1$ *(Cauchy),* $\alpha = 1.5$ *(Holtsmark), and* $\alpha = 2$ *(Gaussian); the lower the index, the sharper and higher the tip of the distribution. All these PDFs were easily obtained by numerical inverse Fourier transform of their characteristic functions* $f(k) = \exp(-|k|^{\alpha})$.

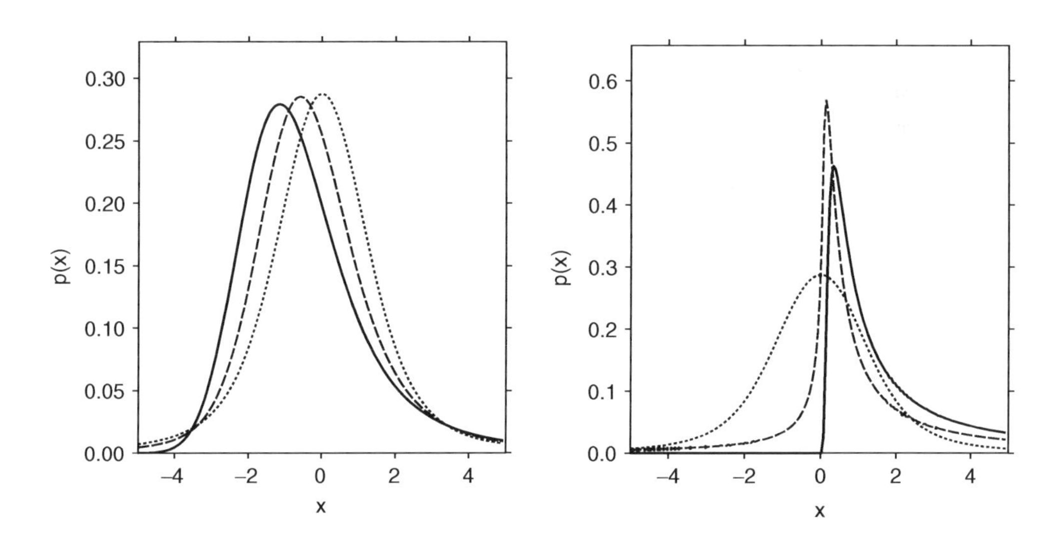

Fig. 7.3 *Skewed Lévy distributions. The left-hand panel shows the members of the "Holtsmark family"* $\alpha = 1.5$. *The dotted line denotes the symmetric distribution* ($\beta = 0$), *the dashed line corresponds to* $\beta = 0.5$, *and the full line indicates* $\beta = 1$. *The right-hand panel shows the members of the "Smirnov family"* $\alpha = 0.5$; *the values of* β *are the same as in the left-hand panel. Note that the distribution for* $\beta = 1$ *is concentrated on the right-hand half-line.*

7.2 Space-fractional diffusion equation for Lévy flights

Lévy flights can be described in the language of fractional diffusion equations. Contrary to heavy-tailed CTRWs, these contain not temporal but spatial fractional derivatives.

Let us concentrate on the case of a symmetric flight. In what follows we use the continuous-time analogy, and simply translate the number of steps n into time t (for example, assuming an exponential or any other narrow distribution of waiting times, so that $t = \tau n$ with τ being the mean time of a step). To obtain the corresponding equation it is enough to remark that according to the discussion in Chapter 1,

$$p_{n+\Delta n}(x) = \int_{-\infty}^{\infty} p_n(y) p_{\Delta n}(x-y) dy$$

holds, where $p_n(y)$ is the PDF of the walker's positions after n steps and $\Delta n << n$. In continuous time the corresponding equation looks like

$$p(x, t+\Delta t) = \int_{-\infty}^{\infty} p(y,t) p(x-y, \Delta t) dy. \tag{7.8}$$

Let us assume that the distribution density of displacement in one step is given by a Lévy distribution, and its characteristic function reads $f(k) = \exp(-\sigma^\alpha |k|^\alpha)$. That of $p_{\Delta n}(x)$ is therefore $f_{\Delta n}(k) = \exp(-\Delta n \sigma^\alpha |k|^\alpha)$, and that of $p_{\Delta t}(x)$ correspondingly reads $f(k, \Delta t) = \exp\left(-\Delta t \frac{\sigma^\alpha}{\tau} |k|^\alpha\right)$. Performing the Fourier transform of Eq.(7.8), we get

$$p(k, t+\Delta t) = p(k,t) f(k, \Delta t) = p(k,t) \exp\left(-\Delta t \frac{\sigma^\alpha}{\tau} |k|^\alpha\right).$$

In the small-k limit, we can expand the exponential, so that

$$p(k, t+\Delta t) = p(k,t) - \Delta t \frac{\sigma^\alpha}{\tau} |k|^\alpha p(k,t),$$

i.e.,

$$\frac{d}{dt} p(k,t) \cong \frac{p(k,t+\Delta t) - p(k,t)}{\Delta t} = -\frac{\sigma^\alpha}{\tau} |k|^\alpha p(k,t). \tag{7.9}$$

The expression on the right-hand side of the equation is nothing other than the Fourier representation of the Riesz (symmetrized Weyl) derivative, which we will here denote by $\frac{d^\alpha}{d|x|^\alpha}$. In coordinate space Eq.(7.9) reads

$$\frac{\partial}{\partial t} p(x,t) \cong \frac{\sigma^\alpha}{\tau} \frac{\partial^\alpha}{\partial |x|^\alpha} p(x,t). \tag{7.10}$$

The combination $\frac{\sigma^\alpha}{\tau}$ plays the role of the generalized diffusion coefficient.

The Riesz–Weyl derivative of a function $f(x)$ is defined for $\alpha \neq 1$ as a sum of two Weyl derivatives,

$$\frac{d^\alpha}{d\left|x\right|^\alpha} f(x) = -\frac{1}{2\cos(\pi\alpha/2)} \left[{}_{-\infty}D_x^\alpha + {}_xD_\infty^\alpha \right] f(x). \tag{7.11}$$

To see this we note that a Fourier transform of ${}_{-\infty}D_x^\alpha f(x)$ reads $(ik)^\alpha f(k)$ (see Chapter 6), while that of ${}_xD_\infty^\alpha f(x)$ is $(-ik)^\alpha f(k)$, so that the combination given by Eq.(7.11) produces under Fourier transform Eq.(7.9).

For $\alpha = 1$ the operator $\frac{d}{d|x|} f(x)$ is related to the Hilbert transform:

$$\frac{d}{d\left|x\right|} = -\frac{d}{dx} \hat{H} f(x) = -\frac{d}{dx} \frac{1}{\pi} \int_{-\infty}^{\infty} \frac{f(x')dx'}{x - x'}.$$

An interesting question arises when the action of external forces on a Lévy walker is considered. The simplest way of introducing the force corresponds to an assumption that this simply induces the drift, which is proportional to the force $f(x,t)$ i.e., to consider jumps and the deterministic displacement as two independent processes running in parallel. In this case the following generalized Fokker–Planck equation for the PDF of the walker's coordinate is obtained:

$$\frac{\partial}{\partial t} p(x,t) \cong K_\alpha \frac{\partial^\alpha}{\partial\left|x\right|^\alpha} p(x,t) - \mu \frac{d}{dx} \left(f(x,t) p(x,t) \right). \tag{7.12}$$

Equation (7.12) can describe some realistic situations when Lévy jumps may be considered as external perturbations of the deterministic system.[3]

Equation (7.12) is not pertinent to systems close to thermodynamic equilibrium. For example, it does not reproduce the Boltzmann distribution as a stationary solution for particles in a confining potential $U(x)$, corresponding to $f(x) = -\operatorname{grad} U(x)$.

This can be easily checked using Exercise 7.4.

Exercise 7.4 Consider a stationary solution to Eq.(7.12) in a parabolic potential $U(x) = \kappa x^2/2$, i.e., the solution to the equation

$$K_\alpha \frac{\partial^\alpha}{\partial\left|x\right|^\alpha} p(x) + \mu\kappa \frac{d}{dx} \left(x p(x) \right) = 0.$$

Show that $p(x)$ is a symmetric Lévy distribution of index α, and find its scaling parameter σ as a function of K_α and κ and μ.

Hint: Fourier transform the equation. Note that under Fourier transform $xf(x)$ corresponds to $-i\frac{d}{dk} f(k)$.

[3] For qualitative investigation of such equations (see e.g. Exercise 7.5) it is useful to remark that for large x the fractional derivative $\frac{\partial^\alpha}{\partial|x|^\alpha} p(x)$ for *any probability density* function $p(x)$ behaves as $|x|^{-1-\alpha}$. To see this pass to the Fourier representation, note that $p(k \to 0) = 1$ and perform the inverse transform.

An interesting property of Eq.(7.12) is that it may possess multimodal stationary solutions (ones with two humps) even in single-well potentials, as exemplified by a quite curious analytical solution for quartic potential and $\alpha = 1$: The solution to the equation

$$\frac{\partial^1}{\partial |x|^1} p(x) + \frac{d}{dx}\left(x^3 p(x)\right) = 0 \tag{7.13}$$

(take care: The Riesz–Weyl fractional derivative of the first order *is not* the first-order derivative!) reads

$$p(x) = \frac{1}{\pi(1 - x^2 + x^4)}.$$

Exercise 7.5 Consider a stationary solution to Eq.(7.12) for the Cauchy–Lévy flight in a quartic potential $U(x) = \kappa x^4/4$ i.e., the solution to the equation

$$K_\alpha \frac{\partial^1}{\partial |x|^1} p(x) + \mu\kappa \frac{d}{dx}\left(x^3 p(x)\right) = 0.$$

Hints: Note that $p(x)$ has to possess the first moment (why?). Put the equation in dimensionless form, Eq.(7.13). Fourier transform the equation. Show that its solution in the Fourier representation reads $p(k) = \frac{2}{\sqrt{3}} \exp(-|k|/2) \cos\left(\sqrt{3}|k|/2 - \pi/6\right)$ and then perform the inverse transform. How do the parameters of $p(x)$ depend on K_α and κ and μ?

The method of introducing force discussed above is not unique, and several other generalizations of the space-fractional diffusion equation have been proposed for different situations. Some are good candidates for describing strange dynamics under normal equilibrium conditions [3].

7.3 Leapover

Let us assume that a one-dimensional Lévy flight x_n starts at $x = 0$ at $n = 0$ and that there is a target located at $x = d > 0$. We are looking for the first-passage "time"

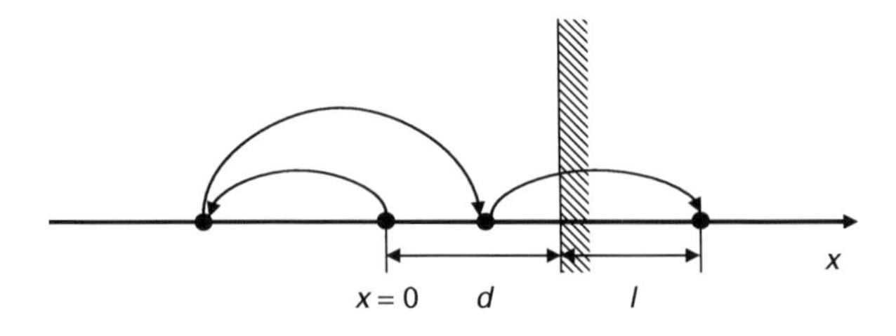

Fig. 7.4 *The definition of leapover.*

$n(d)$ (so that $x_n > d$ and $x_m < d$ for all $0 \le m < n$) and the first-passage leapover $l(d) = x_n - d$ where n is the number of steps to cross or arrive at the target for the first time (see Fig. 7.4).

For the one-sided Lévy flights the first-passage time problem is equivalent to calculating $\chi_n(t)$ in the case of CTRW, now with d playing the role of t. Therefore the result of Exercise 3.7 applies:

$$p_d(n) \approx \frac{d}{\alpha a} n^{-\frac{1}{\alpha}-1} L_\alpha \left(\frac{d}{a\, n^{1/\alpha}} \right)$$

where a is now the characteristic step's length.

Exercise 7.6 Show that in the case $\alpha = 1/2$ the corresponding distribution is a "half-Gaussian."

The first-passage leapover PDF is given by

$$p(l_d) = \frac{\sin(\pi\alpha)}{\pi} \frac{d^\alpha}{l_d^\alpha (d + l_d)} \cong \frac{1}{l_d^{1+\alpha}}$$

for l_d large enough. The PDF of the first-passage leapover is the spatial analog of the forward waiting-time distribution in the CTRW with heavy-tailed waiting times (Eq.(4.8)). In this case the distribution of the leapover length reproduces asymptotically that of the jump length.

The situation for symmetric Lévy flights is somewhat counterintuitive. In this case the first-passage time follows the Sparre Andersen result discussed in Chapter 2, so that $p(n(d)) \propto n^{-3/2}$. The PDF of the leapover length behaves as

$$p(l_d) = \frac{\sin(\pi\alpha/2)}{\pi} \frac{d^{\alpha/2}}{l_d^{\alpha/2}(d + l_d)} \cong \frac{1}{l_d^{1+\alpha/2}},$$

i.e., for large l, as

$$p(l) \propto l^{-1-\alpha/2}$$

(see Ref.[4]). It is interesting to note that the tail of the first-passage leapover decays *more slowly* than the tail of the PDF of the increments of the flight (see Ref.[5]). The method used in Ref.[4] is almost the same as those used in our discussion of Sparre Andersen theorem in Chapter 2. The starting equation in discrete time reads[4]

[4]Ref.[4] uses continuous time and starts from the equation that in our notation reads

$$p(x,t) = p^-(x,t) + \int_0^t \int_d^\infty p^+(y,t')p(x-y,t-t')\,dy\,dt'$$

where the notation is parallel to those in Eq.(7.14) with the only difference being that an argument t is used instead of the index n.

$$p_n(x) = p_n^-(x) + \sum_{m=1}^{n} \int_d^{\infty} p_m^+(y) p_{n-m}(x-y)\, dy. \tag{7.14}$$

The notation follows that used in Chapter 2; here $p_n^-(x)$ is the probability density that a walker who never crossed to the $x > d$-half-line before is at position x after n steps, and $p_n^+(x)$ is the probability density of the positions of the walker who made the first crossing from $x < d$ to $x > d$ at the n-th step. The meaning of the equation is that a particle at x after n steps either never left the half-axis $x < d$, or left it for the first time at step m passing to the position y and then made $n - m$ steps in an arbitrary direction. The leapover distribution is then nothing other than $p(l) = \sum_{m=1}^{\infty} p_m^+(l)$. The analytical methods, again based on the properties of the p^+ and p^- functions, are somewhat more involved than those used in Chapter 2, so we do not reproduce them here.

7.4 Simulation of Lévy distributions

We have seen how important the Lévy distributions are in random walks. Often we want to simulate the Lévy flights exactly, not only as asymptotics of the sums of Pareto-distributed random values. In this section we give the recipes to how this can be done.

The generation of any random variables starts from generating a random variable y distributed homogeneously in the interval [0,1], for which many well-known and tested random-number generators are available [6]. Random variables following different distributions can be easily generated when the corresponding cumulative distribution function $F(x) = \int_{-\infty}^{x} p(x')dx'$ has a known inverse, $F^{-1}(y)$. In this case the variable $x = F^{-1}(y)$ has the PDF $p(x)$. To see this, note the formula of the change of variables in the probability distributions, $p(x) = p(y(x)) \left| \frac{dy}{dx} \right|$, and the fact that $p(y(x)) = 1$ and $dy/dx = \frac{1}{dx/dy} = p(x)$. For example, to generate a variable distributed according to the exponential law, $p(x) = e^{-x}$ (with $F(x) = 1 - e^{-x}$ and $F^{-1}(y) = -\ln(1-y)$), it is enough to generate a homogeneously distributed variable $y \in [0, 1]$ and to take its negative logarithm, $x = -\ln y$ (the variables y and $1 - y$ have the same distribution).

However, the immediate transformation of the homogeneously distributed random variable into the Lévy-distributed one is not a task that can be implemented effectively, since the inverse of the cumulative Lévy function is typically known only in the form of a power series that has to be summed up with considerable accuracy, i.e., up to a considerably high term, and inverted. The effective algorithms for generating Lévy-distributed variables follow another line and are based on the generalization of the famous Box–Muller method for generating a Gaussian.

7.4.1 The Box–Muller method for a Gaussian

The idea of Box and Muller was to consider a simultaneous transformation of the *two* variables x_1 and x_2. Let us consider the two transformed variables

$$
\begin{aligned}
z_1 &= \sqrt{-2\ln x_1}\cos(2\pi x_2) \\
z_2 &= \sqrt{-2\ln x_1}\sin(2\pi x_2)
\end{aligned}. \tag{7.14}
$$

Solving this system of equations for x_1 and x_2 we get

$$
\begin{aligned}
x_1 &= \exp\left[-(z_1^2+z_2^2)/2\right] \\
x_2 &= \frac{1}{2\pi}\arctan\left(\frac{z_2}{z_1}\right)
\end{aligned}.
$$

The variables x_1 and x_2 are defined for all $-\infty < z_1, z_2 < \infty$ and lie in the interval $0 < x_1, x_2 < 1$. The Jacobian of the transformation reads

$$
J = \frac{\partial(x_1, x_2)}{\partial(z_1, z_2)} = \begin{pmatrix} \partial x_1/\partial z_1 & \partial x_1/\partial z_2 \\ \partial x_2/\partial z_1 & \partial x_2/\partial z_2 \end{pmatrix} = -\frac{1}{\sqrt{2\pi}}e^{-z_1^2/2}\cdot\frac{1}{\sqrt{2\pi}}e^{-z_2^2/2},
$$

so that their joint probability density is

$$
p(z_1, z_2) = p\left[x_1(z_1, z_2), x_2(z_1, z_2)\right]\left|\frac{\partial(x_1, x_2)}{\partial(z_1, z_2)}\right| = \frac{1}{\sqrt{2\pi}}e^{-z_1^2/2}\cdot\frac{1}{\sqrt{2\pi}}e^{-z_2^2/2},
$$

which corresponds to that of two independent Gaussian variables provided that the variables x_1 and x_2 are independent and homogeneously distributed, so that $p[x_1(z_1, z_2), x_2(z_1, z_2)] = 1$ for $0 < x_1 < 1,\ 0 < x_2 < 1$ and vanishes elsewhere. Essentially, each of the variables, or both of them, can be used in simulations. We also note that the variable $W = -\ln x_1$ is a non-negative random variable distributed according to an exponential law $p(u) = e^{-u}$. Therefore we can slightly alter the notation and formulate the Box–Muller algorithm as follows:

- Generate a random variable V distributed homogeneously on $(-\pi/2, \pi/2)$.
- Generate an exponential variable W with mean 1.
- Compute $X = \sqrt{2W}\sin V$.

The variable X is distributed according to a Gaussian law with zero mean and unit dispersion.

Exercise 7.7 Show that the random variable $Z = \sqrt{-2\ln x_1}\sin(2\pi x_2)$, with x_1 and x_2 both homogeneously distributed between 0 and 1, and the random variable $X = \sqrt{2W}\sin V$, with V homogeneously distributed on $(-\pi/2, \pi/2)$ and W possessing the probability density $P(W) = e^{-W}$ on $[0, \infty)$, possess the same probability distribution.

7.4.2 Lévy distributions

Lévy-distributed random variables can be obtained through a nonlinear transformation of Gaussian random variables, and thus generated via a method similar to the Box–Muller approach [7–9]. The derivation of this result is based on the integral representation of the corresponding distribution function, which is obtained by performing an inverse Fourier transform of its characteristic function via integration over a cleverly chosen contour in a complex plane, as described by Zolotarev (1983) [10], and is too technical to be discussed here in detail. Therefore we do not give the derivation here and point the reader to the original references for this somewhat cumbersome derivation.

We give here two recipes: the simple one applicable for all symmetric Lévy distributions, and another one for the general case. The last approach is, however, not applicable for asymmetric distributions with $\alpha = 1$, i.e., to all asymmetric members of the Cauchy family.

Symmetric distributions

- Generate a random variable V distributed homogeneously on $(-\pi/2, \pi/2)$.
- Generate an exponential variable W with mean 1.
- Compute

$$X = \frac{\sin(\alpha V)}{[\cos V]^{1/\alpha}} \left\{ \frac{\cos[(1-\alpha)V]}{W} \right\}^{\frac{1-\alpha}{\alpha}}.$$

The variable X is distributed according to a symmetric Lévy law $L_{\alpha,0}(x)$ with zero position parameter and unit scale parameter. This is exactly the algorithm that was used to generate the data for Fig. 7.1.

Asymmetric distributions with $\alpha \neq 1$

- Compute $A = \arctan[\beta \tan(\pi\alpha/2)]$ and the two auxiliary values

$$C = A/\alpha$$
$$D = [\cos A]^{-1/\alpha}.$$

- Generate a random variable V distributed homogeneously on $(-\pi/2, \pi/2)$.
- Generate an exponential variable W with mean 1.
- Compute

$$X = D \frac{\sin[\alpha(V+C)]}{[\cos V]^{1/\alpha}} \left\{ \frac{\cos[V - \alpha(V+C)]}{W} \right\}^{\frac{1-\alpha}{\alpha}}. \quad (7.15)$$

The variable X is distributed according to a Lévy law $L_{\alpha,\beta}(x)$ with zero position parameter and unit scale parameter.

We note that for $\alpha = 1$ the value of A diverges, and therefore the approach does not produce a recipe for generating asymmetric variables in this case. We are not aware of any working method applicable in the case of $\alpha = 1$ and $\beta \neq 0$.

Exercise 7.8 Show that Eq.(7.15) reduces to that of the Box–Muller approach for $\alpha = 2$ and for any β. (Note that the scaling parameter of a Gaussian distribution is $\sqrt{2}$ times smaller than its dispersion!)

References

[1] W. Feller. *An Introduction to Probability Theory and Its Applications*, New York: Wiley 1971 (The corresponding material is discussed in Chapter XVII of the book.)
[2] T. Garoni and N. Frankel. *J. Math. Phys.* **43**, 2670 (2002)
[3] D. Brockmann and I.M. Sokolov. *Chem. Phys.* **284**, 409 (2002)
[4] D. Ray. *Trans. Amer. Math. Soc.* **89**, issue 1, pp. 16–24 (1958)
[5] A.V. Chechkin, R. Metzler, V.Y. Gonchar, J. Klafter, and L.V. Tanatarov. *Journal of Physics A*, Vol. 36, pp. L537–44 (2003)
[6] W.H. Press, S.A. Teukolsky, W.T. Vetterling, and B.P. Flannery. *Numerical Recipes: The Art of Scientific Computing*, 3rd edition, Cambridge: Cambridge University Press, 2007
[7] J.M. Chambers, C.L. Mallows, and B. Stuck. *J. Amer. Statist. Assoc.* **71**, 340–44 (1976)
[8] A. Janicki and A. Weron. *Mathematics and Computers in Simulation* **39**, 9–19 (1995)
[9] R. Weron. *Statistics and Probability Letters*, **28**, 2, 165–71 (1996)
[10] V. Zolotarev. *One-Dimensional Stable distributions*, Providence, RI: American Math. Soc., 1986

Further reading

S. Redner. *A Guide to First-Passage Processes*, Cambridge: Cambridge University Press, 2001

8
Coupled CTRW and Lévy walks

"Time is the longest distance between two places."

Tennessee Williams

As discussed in Chapter 7, Lévy flights constitute an interesting family of superdiffusive random walks whose characteristics are long stretches at all scales. However, physical realizations of such process are quite rare. The reason for this is the divergence of the second moment in the genuine Lévy flight process. Because of this, the application of stable laws in physics was delayed for a long time after their introduction in mathematics. For instance, the process we would now call a Lévy flight with $\alpha = 2/3$ had been considered as a possible model for tracers' dispersion in an isotropic turbulent flow by Monin in 1955 (see Ref.[1]). However, by this time it was already clear that the corresponding theory did not adequately describe large-scale dispersion, because of divergence. In the present chapter we discuss ways to remedy the divergence of the second moment by coupling spatial and temporal aspects of the walk by connecting the length of a stretch to its time cost. This gives rise to a class of space–time coupled processes, Lévy walks being one of the most prominent examples. In this chapter we use the one-dimensional notation, but the coordinate x can be considered as a vector as well.

8.1 Space–time coupled CTRWs

As discussed in Chapter 1, random walks with a PDF for step lengths that is a stable law with a heavy tail

$$p(x) \propto \frac{1}{|x|^{1+\alpha}}, \quad 0 < \alpha < 2 \tag{8.1}$$

have a diverging second moment $\langle x^2 \rangle \to \infty$. These simple random walks do not explicitly take into account time. This total neglect of temporal aspects is a drawback of such description in applications. We can introduce the temporal aspect in a way that the spatial and the temporal contributions are decoupled, $\psi(x,t) = p(x)\psi(t)$, which, following Eq.(3.10) in Chapter 3, leads in Fourier–Laplace space to

$$P(k,s) = \frac{1-\psi(s)}{s} \frac{1}{1-p(k)\psi(s)}. \tag{8.2}$$

Taking $p(x)$ as given by Eq.(8.1) and $\psi(t) \propto t^{-1-\beta}$ with $0 < \beta < 1$ corresponding to $p(k) \propto 1 - |k|^{\alpha}$ and $\psi(s) \propto 1 - s^{\beta}$, we get

$$P(k, s) = \frac{s^{\beta-1}}{s^{\beta} + k^{\alpha}}$$

and see that the second moment still diverges. Thus in the case where the "time cost" of a step does not depend on the step length, the power-law behavior of step lengths dominates, so that the MSD at any given time t is infinite.

The divergence in the MSD can be overcome when longer steps are penalized by higher time cost that introduces space–time coupling into the PDF

$$\psi(x, t) = p(x) f(t|x), \tag{8.3}$$

where $f(t|x)$ is a conditional probability that a step of length x requires time t for its completion [2, 3].

There is an equivalent way of putting Eq.(8.3) that emphasizes not the distances traversed but the times needed to go from one turning point to another one. We can specifically write

$$\psi(x, t) = \psi(t) \phi(x|t), \tag{8.4}$$

with $\psi(t)$ being the PDF of times required to perform a step, and $\phi(x|t)$ is a conditional probability that a step requiring time t for completion has a length x.

Our aim now is to obtain the equation for the PDF of the walker's position under correlated random walks. Let $\eta(x, t)$ be the PDF of the positions of walkers *just arriving* at x at time t after completing a step. This position is given by the usual recurrent relation

$$\eta(x, t) = \int dx' \int dt' \eta(x', t') \psi(x - x', t - t') + \delta(t)\delta(x). \tag{8.5}$$

The first term on the right-hand side of this equation corresponds to the case in which the last completed step was that to the point x' finished at some $t' < t$, and the actual step of the walker was that from x' to the actual position x and took time $t - t'$ for its completion. The second term corresponds to the initial condition, i.e., to the fact that the actual step can be the zeroth one. In Fourier–Laplace representation Eq.(8.5) takes an especially simple form

$$\eta(k, s) = \eta(k, s) \psi(k, s) + 1,$$

leading to the solution

$$\eta(k, s) = \frac{1}{1 - \psi(k, s)}. \tag{8.6}$$

Let us now consider the position x at time t of the random walker who arrived at y with its last step completed at time $t_1 < t$. Let $\Psi(x, t)$ be the PDF characterizing the

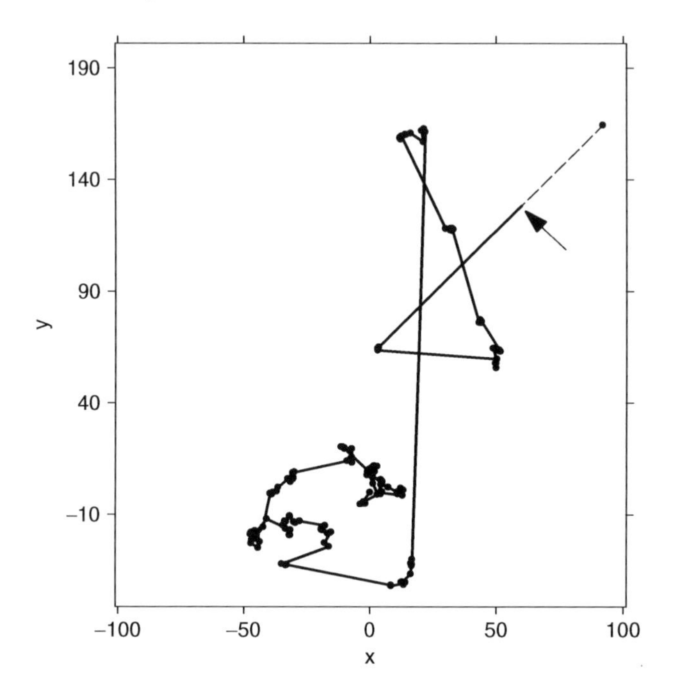

Fig. 8.1 *The trajectory of the Lévy walk (full line) and its turning points (dots). The distance between turning points is distributed according to a heavy-tailed distribution (we use Cauchy distribution here). The trajectory of the walk is terminated at time t before reaching the last turning point. The terminal point of the trajectory is indicated by the arrow.*

displacement of the random walker during the last uncompleted step (e.g., along the last straight line in Fig. 8.1).

The PDF of the walker's position at time t is then

$$P(x,t) = \int dx' \int dt' \eta(x',t')\Psi(x-x',t-t')$$

or, in Fourier–Laplace representation,

$$P(k,s) = \eta(k,s)\Psi(k,s). \tag{8.7}$$

Combining Eqs.(8.6) and (8.7), we get

$$P(k,s) = \frac{\Psi(k,s)}{1-\psi(k,s)}. \tag{8.8}$$

The form of $\Psi(x,t)$ depends on the particular process considered.

As a simplest example let us consider the situation where we still have jumps taking place instantaneously that are separated by rest periods whose duration is given by a waiting-time distribution $\psi(t) = \int_{-\infty}^{\infty} \psi(x,t)\,dx$. The joint probability distribution of step lengths and times is given by $\psi(x,t) = \psi(t)\phi(x|t)$. In this case, the last, incomplete, step corresponds essentially to the rest period of the particle, so that

$$\Psi(x,t) = \delta(x) \int_t^\infty \psi(t')dt' \tag{8.9}$$

and

$$\Psi(k,s) = \Psi(s) = \frac{1-\psi(s)}{s}, \tag{8.10}$$

for which

$$P(k,s) = \frac{1-\psi(s)}{s} \frac{1}{1-\psi(k,s)}. \tag{8.11}$$

The main result for the time–space coupled random walks is given by the Fourier–Laplace transform of the PDF to find a walker at position $\mathbf{r}$ at time t:

$$P(\mathbf{k},u) = \frac{\Psi(\mathbf{k},s)}{1-\psi(\mathbf{k},s)}$$

Here $\psi(\mathbf{x},t)$ is the joint probability of the step's length and duration, and $\Psi(\mathbf{x},t)$ is the PDF characterizing the displacement during the last, uncompleted, step.

Example 8.1 Let us consider an example in which $\psi(t)$ is given by $\psi(t) \propto t^{-1-\beta}$ as before (so that its Laplace transform for small s is given by $\psi(s) \cong 1 - as^\beta$), and the step's length is proportional to the preceding waiting time: $\phi(x|t) = \frac{1}{2}[\delta(x-ct) + \delta(x+ct)]$ where c is a constant having a dimension of velocity, so that

$$\psi(x,t) \propto \frac{1}{2}\delta(|x| - ct)\psi(t). \tag{8.12}$$

The Fourier–Laplace transform of $\psi(x,t)$ is given by

$$\begin{aligned} \psi(k,s) &= \frac{1}{2}\int e^{ikx-st}\left[\delta(-x-ct) + \delta(x-ct)\right]\psi(t)dxdt \\ &= \frac{1}{2}\left[\int e^{-(s+ick)t}\psi(t)dt + \int e^{-(s-ick)t}\psi(t)dt\right] \\ &= \frac{1}{2}\left[\tilde{\psi}(s+ick) + \tilde{\psi}(s-ick)\right] \equiv Re\tilde{\psi}(s+ick). \end{aligned} \tag{8.13}$$

Let us calculate the MSD of a walker in such a process. For this we are interested in the behavior of $P(k,s)$ for s fixed and $k \to 0$, so that

$$\psi(k,s) \cong 1 - \frac{1}{2}\left[a(s+ick)^\beta + a(s-ick)^\beta\right] = 1 - as^\beta - \frac{ac^2}{2}\beta(1-\beta)k^2 s^{\beta-2}.$$

Substituting this into Eq.(8.11) we get

$$P(k,s) = \frac{s}{s^2 + b^2k^2} \tag{8.14}$$

where $b^2 = \beta(1-\beta)c^2/2$ includes constants in the problem. The MSD is then given by $\langle x^2(t)\rangle \cong b^2t^2$, which corresponds to a ballistic behavior, although emerging from a random-walk process. Moreover, the value of b^2 is always smaller than c^2. The inverse Fourier–Laplace transform of Eq.(8.14) can be easily taken and leads to the coarse-grained form for the PDF of the walker's positions

$$P(x,t) = \frac{1}{2}\left[\delta(x-bt) + \delta(x+bt)\right],$$

which corresponds to two waves running in opposite directions with the velocity b rather than c.

Exercise 8.1 Calculate the Fourier–Laplace representation $\psi(k,s)$ for Lévy walk $\psi(x,t) \propto \delta(|x| - t^\nu)\psi(t)$ with $\psi(t) \propto t^{-1-\alpha}$ for $k \to 0$ and for s small, for both cases $0 < \alpha < 1$ and $1 < \alpha < 2$ and for different relations between ν and α.

Hint: Change to the form Eq.(8.3).

Example 8.2 As another example let us consider the space–time coupled random walk with

$$\phi(x|t) = \frac{1}{\sqrt{2\pi D(t)}} \exp\left(-\frac{x^2}{2D(t)}\right),$$

and take $D(t) \propto t^m$ with $m > \alpha$. In this case the MSD behaves as

$$\langle x^2(t)\rangle \propto t^m. \tag{8.15}$$

Exercise 8.2 Derive the result Eq.(8.15).

Hint: Obtain the form for $\Psi(k,s)$ first.

8.2 Lévy walks

Let us now turn to a closely related but different model in which the walker moves with a constant velocity v between the turning points of its trajectory (see Fig. 8.1).

In what follows, we consider a one-dimensional situation as depicted in Fig. 8.2. Parallel to the previous case, we have

$$\psi(x,t) \propto \frac{1}{2}\delta(|x| - vt)\psi(t) \tag{8.16}$$

where $\psi(t)$ is now the PDF of times spent in homogeneous rectilinear motion. Let us, moreover, assume $\psi(t)$ follows a power law $\psi(t) \propto t^{-1-\alpha}$, now with $0 < \alpha < 2$. This process can be visualized as the motion of a walker with a constant velocity between

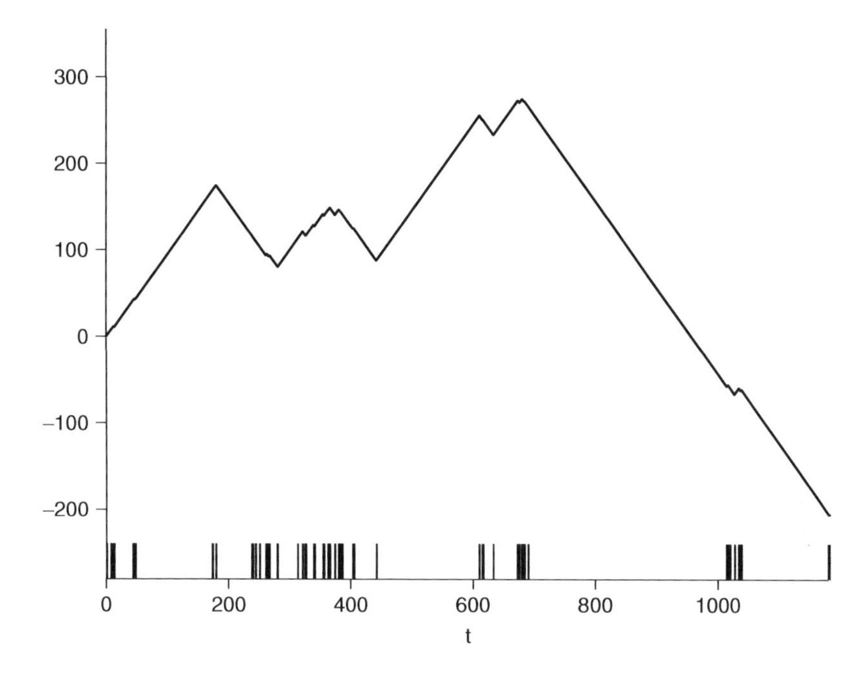

Fig. 8.2 *A trajectory of a one-dimensional Lévy walk corresponding to exactly the same realization of waiting times as the CTRW in Fig. 3.3 (barcode in the lower part of the plot).*

the turning points visited by a Lévy flight, the fractal set of points of fractal dimension α discussed at the beginning of this chapter.

Since the expression for $\psi(x,t)$ (at least for $0 < \alpha < 1$) is the same as in the previous case, the same techniques to obtain $\psi(k,s)$ can be used. The difference between the two processes stems from the difference in $\Psi(x,t)$, which is now given by

$$\Psi(x,t) = \frac{1}{2}\delta(|x|-t)\int_t^{\infty}\psi(t')dt' \tag{8.17}$$

(compare with Eq.(8.9)), which leads to a more complicated form for $P(k,s)$. Parallel to the steps leading to Eq.(8.13), $\Psi(k,s)$ can be written as

$$\Psi(k,s) = Re\tilde{\Psi}(s+ivk) \tag{8.18}$$

with $\tilde{\Psi}(s) = [1-\psi(s)]/s$.

Before turning to the genuine Lévy walks with power-law waiting-time distributions, let us consider the situation with $\psi(t)$ following the exponential law, which often appears as an example of motion showing a crossover between the ballistic and the diffusive behavior:

Exercise 8.3 Consider a space–time coupled random walk as governed by Eqs.(8.16) and (8.17) with exponentially distributed waiting times $\psi(t) = \tau^{-1}\exp(-t/\tau)$. Show that its PDF fulfills the *telegrapher's equation*

$$c^2 \frac{\partial^2}{\partial x^2} P(x,t) = \frac{\partial^2}{\partial t^2} P(x,t) + \gamma \frac{\partial}{\partial t} P(x,t).$$

Find the values of the parameters c and γ and define the corresponding initial condition.

Hint: Compare the result for the Fourier–Laplace transformed probability density, Eq.(8.8), with the general solution to the telegrapher's equation in the Fourier–Laplace domain.

In considering Lévy walks, we have to distinguish between the two cases $0 < \alpha < 1$ and $1 < \alpha < 2$. Let us first concentrate on the MSD of the random walker. For $0 < \alpha < 1$ for s fixed and $k \to 0$, we get

$$P(k,s) = \frac{1}{s} \frac{s^\alpha - d_1^2 k^2 s^{\alpha-2}}{s^\alpha + b_1^2 k^2 s^{\alpha-2}} \tag{8.19}$$

with $b_1^2 = \alpha(1-\alpha)v^2/2$ and $d_1^2 = (\alpha-1)(\alpha-2)v^2/2$, which gives the ballistic behavior

$$\langle x^2(t)\rangle \cong V^2 t^2, \tag{8.20}$$

The Lévy walk is a space–time coupled random-walk model characterized by

$$\psi(x,t) \propto \frac{1}{2}\delta(|x| - vt)\psi(t)$$

and

$$\Psi(x,t) = \frac{1}{2}\delta(|x| - t)\int_t^\infty \psi(t')dt'$$

with $\psi(t)$ following a power law $\psi(t) \propto t^{-1-\alpha}$ with $0 < \alpha \leq 2$.

with the characteristic spread velocity given by $V^2 = (1-\alpha)v^2 < v^2$.

The behavior for $1 < \alpha < 2$ is vastly different and more interesting. For $1 < \alpha < 2$ the waiting-time density $\psi(t)$ possesses the first moment $\tau_1 = \langle t\rangle = \int_0^\infty t\psi(t)dt$, so that $\psi(s) \cong 1 - \tau_1 s - c_1 s^\alpha$ with c_1 being a constant. Now

$$\psi(k,s) \cong 1 - \tau_1 s - \frac{1}{2}\left[c_1(s+ivk)^\alpha + c_1(s-ivk)^\alpha\right] = 1 - \tau_1 s - c_1 s^\alpha - \frac{c_1 v^2}{2}\alpha(\alpha-1)k^2 s^{\alpha-2}. \tag{8.21}$$

For small s the third term can be neglected compared to the second one. For $\Psi(k,s)$ we obtain

$$\Psi(k,s) \cong \tau_1 + \frac{1}{2}\left[c_1(s+ick)^{\alpha-1} + c_1(s-ick)^{\alpha-1}\right]$$
$$= \tau_1 + c_1 s^{\alpha-1} - \frac{c_1 v^2}{2}(\alpha-1)(\alpha-2)k^2 s^{\alpha-3}. \tag{8.22}$$

Here the second term can be neglected compared to the first one. We thus have

$$P(k,s) = \frac{1}{s}\frac{\tau_1 + d_2^2 k^2 s^{\alpha-2}}{\tau_1 + b_2^2 k^2 s^{\alpha-2}}, \tag{8.23}$$

giving us for the MSD

$$\left\langle x^2(t)\right\rangle = const \cdot t^{3-\alpha}. \tag{8.24}$$

This result interpolates between the normal diffusive case for $\alpha > 2$ and the ballistic case for $\alpha < 1$, as discussed before. The marginal cases $\alpha = 1$ and $\alpha = 2$ give rise to logarithmic corrections (see the Box).

Exercise 8.4 For $\alpha > 2$ the Lévy walk process merges with normal diffusion. Show that for $\psi(t) \propto t^{-1-\alpha}$ with $\alpha > 2$ the MSD behaves as $\langle x^2(t)\rangle \propto t$.

Exercise 8.5 Calculate the MSD in a Lévy walk with $\alpha = 1$ and with $\alpha = 2$. Show that for these cases logarithmic corrections appear.

The MSD in Lévy walks follows the pattern

$$\left\langle x^2(t)\right\rangle \propto \begin{cases} t^2, & 0 < \alpha < 1 \\ t^2/\ln t & \alpha = 1 \\ t^{3-\alpha} & 1 < \alpha < 2 \\ t\ln t & \alpha = 2 \\ t & 2 < \alpha \end{cases}$$

We note that the approximation used (namely, keeping s constant, and expansion for small k) leading to the correct results for the MSD, depending only on the form of $P(k,s)$ for $k \to 0$, is not sufficient to obtain the full form of the propagator, especially in its central part. Obtaining such forms is a much harder task.

The forms of the propagators $P(x,t)$ also differ vastly in the cases $0 < \alpha < 1$ and $1 < \alpha < 2$. In both cases, of course, $P(x,t)$ vanish outside the interval $-vt \le x \le vt$. Inside this interval the behavior for $0 < \alpha < 1$ corresponds to a PDF with a minimum in the center and (integrable) singularities at the ends of the interval. This behavior is exemplified by an analytically known form corresponding to $\alpha = 1/2$, the so-called arcsinus law

$$P(x,t) = \frac{1}{\pi}\frac{1}{\sqrt{v^2t^2 - x^2}} \tag{8.25}$$

(the term "arcsinus law" refers to the form of the corresponding cumulative distribution function $F(x,t) = \int_{-vt}^{x} P(x,t)dx$).

To obtain Eq.(8.25), we need an expansion for $P(k,s)$ different to that given by Eq.(8.19): We are no longer allowed to assume that $vk << s$. From Eqs.(8.8) and (8.18), we now get

$$P(k,s) = \frac{(s+ivk)^{-1/2} + (s-ivk)^{-1/2}}{(s+ivk)^{1/2} + (s-ivk)^{1/2}} = \frac{1}{\sqrt{s^2+v^2k^2}}. \tag{8.26}$$

Performing the inverse Fourier transform, we get

$$P(x,s) = \frac{1}{2v}K_0\left(\frac{x}{v}s\right), \tag{8.27}$$

where $K_0(z)$ is the modified Bessel function. The Laplace inversion then gives Eq.(8.25).

The overall behavior of the distribution can be described as follows. Let us introduce the rescaled variables and measure time in units of τ_1, so that $\tilde{t} = t/\tau_1$, and lengths in the units of $v\tau_1$, so that $\tilde{x} = x/v\tau_1$. We can then distinguish three parts of the distribution:

- In its central part, $-\tilde{t}^{1/\alpha} < \tilde{x} < \tilde{t}^{1/\alpha}$ the behavior is Gaussian, $P(\tilde{x},\tilde{t}) \cong \tilde{t}^{-1/\alpha}\exp\left[-const\cdot\frac{\tilde{x}^2}{\tilde{t}^{2/\alpha}}\right]$, corresponding to the contribution of the particles changing the direction of their motion many times prior to observation.
- At the flanks the distribution follows a power-law typical of Lévy flights: For $\tilde{t}^{1/\alpha} < |\tilde{x}| < \tilde{t}$ it follows asymptotically the law $P(\tilde{x},\tilde{t}) \cong \tilde{t}/|\tilde{x}|^{1+\alpha}$.
- Exactly at $|\tilde{x}| = \tilde{t}$ it possesses two δ-functional "horns" (called *chubchiks*) of decaying weight, $P(\tilde{x},\tilde{t}) \cong \tilde{t}^{1-\alpha}\delta(|\tilde{x}| - \tilde{t})$, representing the contribution of the particles that moved rectilinearly all the time and never changed their direction. These δ-functions are therefore analogs of the δ-function in the center of the distribution of particle displacements in ordinary CTRW, which represents the contribution of the particles that did not move until time t.

Example 8.3 Transport in a splitting flow: Imagine a flow of fluid between two plates moving in opposite directions (a shear flow). If we have to work with a Newtonian fluid (such as water) the velocity profile will be linear (see Fig. 8.3).

If we shear ketchup, the profile is very different. Many complex fluids such as ketchup show shear thinning: Their viscosity decays under strong velocity gradients. The velocity profile in ketchup looks somewhat like that shown in the middle panel of Fig. 8.3. Let us approximate this profile by a step function, so that the flow velocity is

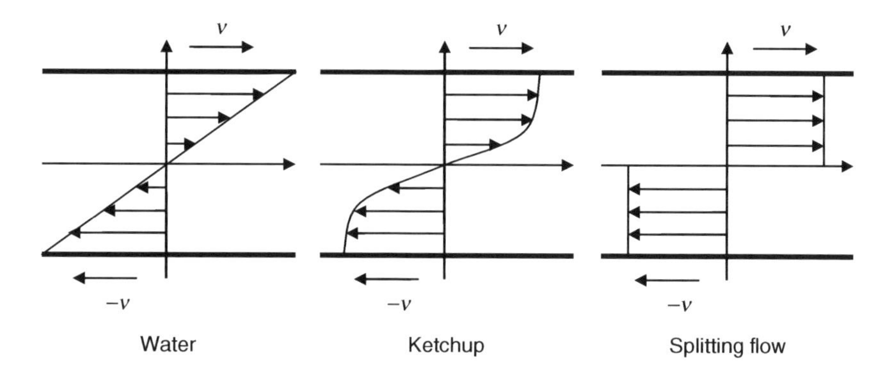

Fig. 8.3 *Velocity profiles in different model fluids. The splitting flow depicted in the right-hand panel corresponds to a Lévy walk in the horizontal direction for a particle diffusing in the vertical direction.*

$v = v_0$ for $y > 0$ and $v = -v_0$ for $y < 0$. Consider now a particle that performs random walks (Brownian motion) with respect to the moving fluid. This particle is carried to the right when in the upper part of the flow, and to the left when it stays in the lower part. For the sake of simplicity, imagine that at time $t = 0$ it starts at $x = y = 0$. At longer times we can neglect the diffusive motion in the x-direction compared to deterministic drift. During times between crossing the zero line, the particle moves with the constant velocity $v = \pm v_0$. The durations of periods of rectilinear motion are given by the return times to $y = 0$, which correspond to $\psi(t) \propto t^{-3/2}$, as discussed in Chapter 2. The situation considered thus exactly corresponds to a Lévy walk with $\alpha = 1/2$.

8.3 Lévy walk interrupted by rests

We can imagine situations where the stretches of a Lévy walk process are interrupted by periods of immobilization at the turning points. A physical example of such a process is discussed at the end of this section. This new process necessitates combining a CTRW process from Chapter 3 with Lévy walks. Parallel to $\psi(x,t)$ and $\Psi(x,t)$ (Eqs.(8.16) and (8.17)) characterizing a Lévy walk process, we have to introduce the distribution of rest times at turning points, given by the PDF $\psi_r(t)$. The corresponding probability of not leaving a turning point up to time t (parallel to Eq.(8.10)) is given by $\Psi_r(t) = \int_t^\infty \psi_r(t')dt'$. Since the periods of resting and moving with a constant velocity alternate, we have

$$\begin{aligned} P(x,t) &= \Psi(x,t) + \int_0^t \psi(x,t')\Psi_r(t-t')dt' \\ &+ \int_{-\infty}^{\infty} dx' \int_0^{\infty} dt' \int_0^{t'} dt''\psi(x',t'')\psi_r(t'-t'')\Psi(x-x',t-t') + \dots \end{aligned} \tag{8.28}$$

where the first term denotes the probability of reaching location x in time t in a single motion event. The second term is the probability of reaching x at an earlier time and to rest until time t. The third term is the probability of reaching x in two periods of motion interrupted by one rest. Taking the Fourier–Laplace transform and summing separately over the even and odd terms (corresponding to events ending by a period of rest or a period of motion), we obtain

$$P(k,s) = \frac{\Psi(k,s) + \Psi_r(s)\psi(k,s)}{1 - \psi_r(s)\psi(k,s)}. \tag{8.29}$$

Let us consider the situation where the corresponding waiting-time PDFs are given by $\psi(t) \propto t^{-1-\alpha}$ and $\psi_r(t) \propto t^{-1-\gamma}$. The result for the MSD reads $\langle x^2(t)\rangle \propto t^{\zeta}$ with

$$\zeta = \begin{cases} 2 + \min\{\gamma, \alpha\} - \alpha, & 0 < \alpha < 1 \\ 2 + \min\{\gamma, 1\} - \min\{2, \alpha\}, & 1 < \alpha \end{cases}. \tag{8.30}$$

For $1 < \alpha < 2$, we have in particular

$$\langle x^2(t)\rangle \propto \begin{cases} t^{2+\gamma-\alpha}, & 0 < \gamma < 1 \\ t^{3-\alpha}, & 1 < \gamma \end{cases}. \tag{8.31}$$

In the first case ($0 < \gamma < 1$) we note that the competition between the Lévy walk stretches and local rests can lead to the compensation of the two effects, i.e., to a MSD following the simple diffusive behavior. In the second case we recover the result for Lévy walks without rests, which affect only the prefactor and not the scaling of the MSD.

The expressions above closely describe the experiments on the flow in a rotating annulus as probed by tracer particles [4]. The entire setup rotates as a rigid body; the flow is generated by pumping fluid through the holes on the bottom. The flow consists of vortices and jets. Figure 8.4 (left panel) shows the trajectories of several tracers in such a flow under different conditions [5]. Individual particles alternately are trapped inside a vortex, and travel long distances in the jet. The angular coordinate of the particle is recorded so that the output of the experiment is the one-dimensional walk consisting of long stretches in jets and rests in vertices (compare to the scheme in Fig. 8.3).

The results of these experiments are well described by the second case in Eq.(8.31).

Hamiltonian systems provide another physical situation where Lévy walks naturally emerge. An example leading to the Lévy walks with rests (again corresponding to the second case in Eq.(8.31), now in two dimensions) is the frictionless motion of a particle in an "egg-crate" potential

$$V(x,y) = A + B(\cos x + \sin y) + C \cos x \cos y, \tag{8.32}$$

as discussed in Ref.[4]. This Hamiltonian model (with the parameters $A = 2.5$, $B = 1.5$, and $C = 0.5$) leads to enhanced diffusion for the values of energies E between

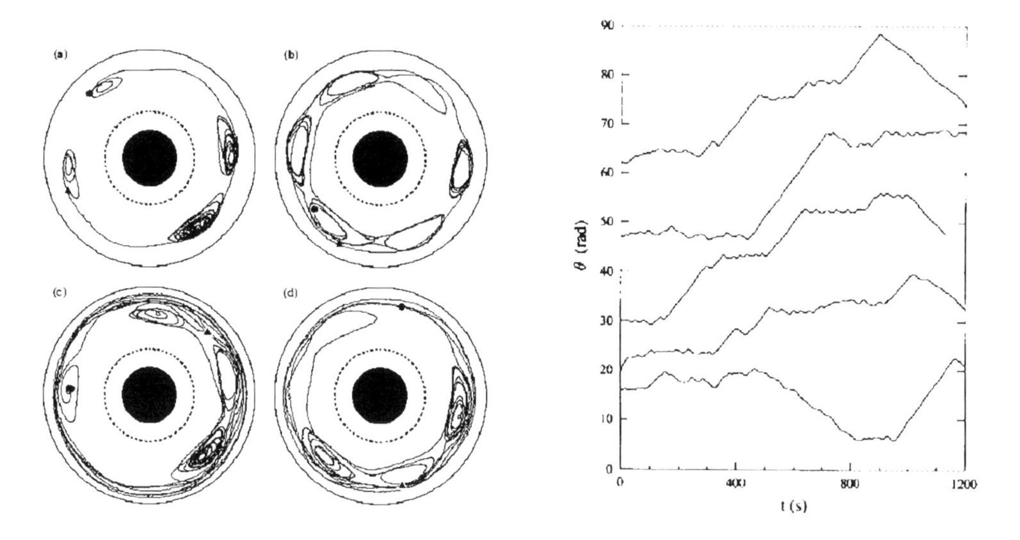

Fig. 8.4 *Left panel: Some typical particle trajectories. The beginning of each trajectory is marked with a circle, the end with a triangle. Right panel: The angular coordinate of a particle as a function of time for five trajectories (cf. Fig. 8.2). The oscillations of the tracer particles correspond to trapping in the vortices, and the diagonal lines correspond to flights [5].*

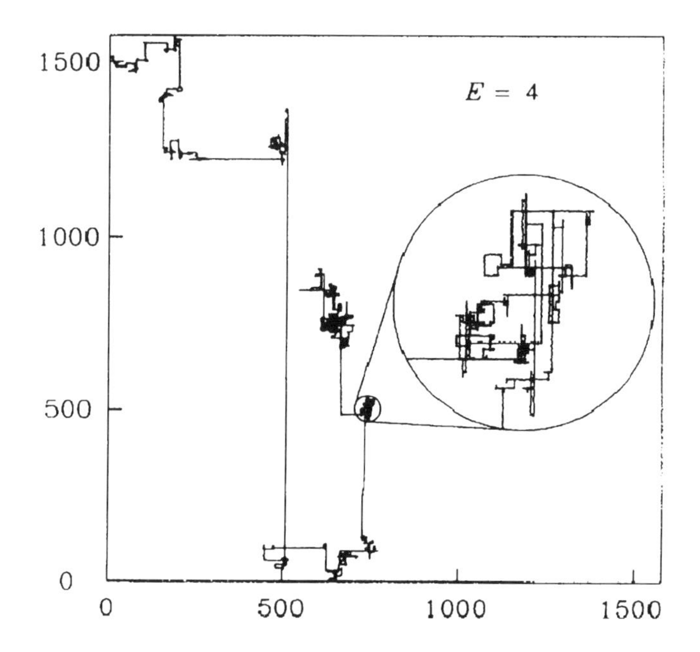

Fig. 8.5 *A typical two-dimensional trajectory* $\mathbf{r}(t)$ *for a particle motion in an "egg-crate" potential (Eq.(8.32)), for* $t = 10^5$.

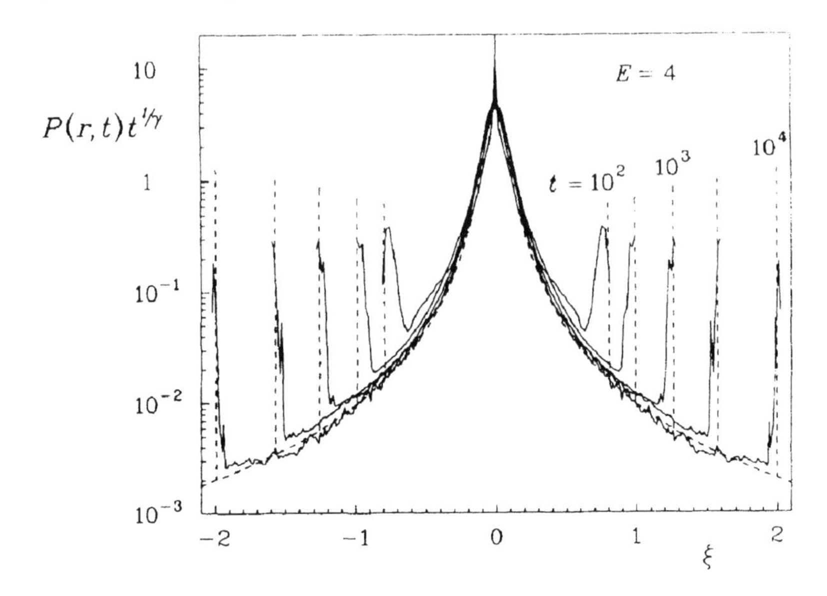

Fig. 8.6 *Simulation results for the particle's motion in an "egg-crate" potential. The dashed lines correspond to the Lévy flight-like behavior in the intermediate part of the distribution and to delta peaks [6].*

2 and 4.5. A typical trajectory is shown in Fig. 8.5. The form of the PDF of particles' displacement in this case is shown in Fig. 8.6, borrowed from Ref.[6].

Hamiltonian systems provide a wealth of examples of trajectories in a phase space that follow Lévy walks in the range of parameters where the system is on the edge of chaos [7]. We also note that some of the features of the Lévy walk behavior carry over to systems with friction [8], where the corresponding behavior follows as intermediate asymptotics.

References

[1] A.S. Monin and A.M. Yaglom. *Statistical Fluid Mechanics*, Vol. 2, Cambridge, MA: MIT Press, 1975

[2] M.F. Shlesinger, J. Klafter, and Y.M. Wong. *Journ. Stat. Phys.* **27**, 499 (1982)

[3] M.F. Shlesinger, B.J. West, and J. Klafter. *Phys. Rev. Lett.* **58**, 1100 (1987)

[4] T.H. Solomon, E.R. Weeks, and H.L. Swinney. *Phys. Rev. Lett.* **71**, 3975 (1993)

[5] T.H. Solomon, E.R. Weeks, and H.L. Swinney. *Physica D* **76**, 70 (1994)

[6] J. Klafter and G. Zumofen. *Phys. Rev. E* **49**, 4873 (1994)

[7] G.M. Zaslavsky. *Phys. Reports* **371**, 461 (2002)

[8] J.M. Sancho, A.M. Lacasta, K. Lindenberg, I.M. Sokolov, and A.H. Romero. *Phys. Rev. Lett.* **92**, 250601 (2004)

9

Simple reactions: $A + B \to B$

"He who seeks does not find, but he who does not seek will be found."

Franz Kafka

In Chapter 2, discussing the number of distinct sites visited, we stated the role this quantity plays, e.g., in chemical kinetics. Now it is time to discuss this issue in some detail. Whatever the mechanism of the chemical reaction, the reactions can occur only when the reacting particles are close to each other, to allow for interaction, e.g., due to quantum-mechanical tunneling. Such a reaction between immobile particles (reaction due to "direct transfer") will be considered in Sec. 9.1 and constitutes the simplest case. In more complex cases, as often takes place under low reactant concentrations, the reactants have first to be brought close together to interact. Diffusion, i.e., random walk of reactants, is often the mechanism that allows for such a reaction. In what follows we consider the simplest reaction $A + B \to B$, the catalytic quenching of excitations, as often observed, e.g., in luminescence. A is essentially an excited state of a molecule, and the excitation can jump from one molecule to another, until a defect or a molecule of the other type, immobile B, is met. Having arrived at B, the excitation dies (is quenched), for example, emitting a photon. Such a situation is often referred to as trapping. In other situations it can also happen that A is immobile and B moves. Surprisingly enough, these two situations are vastly different, and the first one cannot be adequately described in the language of reaction–diffusion equations so popular in modern chemistry and physics. The problem is that this reaction is *fluctuation dominated.* Fluctuation-induced effects are known also for other simple reactions: $A + B \to 0$, $A + B \to 2B$, etc. Such effects are connected with statistical distributions of particles and are eradicated by mixing, which is not a secondary procedure, but is of primary importance for establishing the "classical kinetics" of chemical reactions. The question of mixing lies, however, far beyond the scope of this book.

9.1 Configurational averaging

We start with an example in which the decay of the survival probability is not due to diffusion. Thus, let us consider a random distribution of occupied sites on the lattice that we now call "acceptors," and one special site (at the origin, $\mathbf{r} = 0$) occupied by the "donor." At $t = 0$ the donor is in an excited state and can lose its excitation by transferring it directly to one of the acceptors at $\mathbf{r}_i$. The rate of such transfer depends on the distance between the donor and the acceptor and is given by $w(\mathbf{r}_i,\ \mathbf{0})$. The reverse transport is forbidden. The probability that the excitation *is not* transferred to an acceptor at $\mathbf{r}_i$ up to time t is then given by

$$f(t, \mathbf{r}_i, \mathbf{0}) = e^{-tw(\mathbf{r}_i, 0)},$$

and the overall survival probability of a donor in an excited state up to time t is given by

$$\Phi(t, \mathbf{r}_0) = \prod_{\mathbf{r}_i \neq 0} \left\{ \sum_j g(j) \left[f(t, \mathbf{r}_i, \mathbf{0}) \right]^j \right\},$$

where $g(j)$ is the probability of finding j acceptors on a given site. This is an exact expression.

As an example let us consider the Poisson distribution

$$g(j) = \frac{e^{-p} p^j}{j!}$$

$$\Phi(t) = \prod_{\mathbf{r}_i \neq 0} \left\{ \sum_j \frac{e^{-p} p^j}{j!} \left[f(t, \mathbf{r}_i, \mathbf{0}) \right]^j \right\} = \prod_{\mathbf{r}_i \neq 0} \left\{ \sum_j \frac{e^{-p} [p f(t, \mathbf{r}_i, 0)]^j}{j!} \right\}$$

$$= \prod_{\mathbf{r}_i \neq 0} \left\{ e^{-p} e^{p f(t, \mathbf{r}_i, 0)} \right\} = e^{-\sum_{\mathbf{r}_i \neq 0} p}\, e^{p \sum_{\mathbf{r}_i \neq 0} f(t, \mathbf{r}_i, 0)} = e^{-p \sum_{\mathbf{r}_i \neq 0} [1 - f(t, \mathbf{r}_i, 0)]}. \tag{9.1}$$

$$= e^{-p \sum_{\mathbf{r}_i \neq 0} [1 - f(t, \mathbf{r}_i)]}$$

As a second, and more realistic, example, let us consider a binary situation where each site can be occupied by an acceptor with probability p and may be empty with the probability $1 - p$,

$$g(j) = (1 - p)\delta_{0,j} + p\delta_{1,j}$$

$$\Phi(t, \mathbf{r}_0) = \prod_{\mathbf{r}_i \neq \mathbf{r}_0} \left\{ 1 - p + p e^{-tw(\mathbf{r}_i)} \right\}. \tag{9.2}$$

Exercise 9.1 Show that for $p << 1$ the survival probability $\Phi(t, \mathbf{r}_0)$, Eq.(9.2), tends to $\Phi(t,\ \mathbf{r}_0) = \exp\left\{ -p \sum_{\mathbf{r}_i \neq \mathbf{r}_0} [1 - \exp(-tw(\mathbf{r}_i))] \right\}$, which is exactly the result (9.1).

Hint: Take the logarithm of both sides of Eq.(9.2).

9.2 A target problem

Now let us consider a target problem, where no direct transfer takes place, but the excitation is quenched when one of the acceptors (which are now mobile and perform independent random walks on the lattice) arrives at the donor's site (target). Let us denote by $F_m(\mathbf{r})$ the probability of the first arrival at the origin at the m-th step while starting at site $\mathbf{r}$. Let us moreover define

$$H_n(\mathbf{r}) = \sum_{m=1}^{n} F_m(\mathbf{r}) \tag{9.3}$$

as the probability of arriving at the origin within the first n steps. The probability of not arriving at the donor's site (i.e., survival) is

$$\phi_n(\mathbf{r}) = 1 - H_n(\mathbf{r}).$$

Following similar steps to those in Sec. 9.1, the survival probability of the donor at the origin after n steps is given by

$$\Phi_n = \prod_{\mathbf{r}_i \neq 0} \left\{ \sum_j g(j) \left[\phi_n(\mathbf{r})\right]^j \right\}.$$

Taking $g(j)$ to be Poissonian (which, as we saw, approximates the survival probability in the case of binary disorder) we obtain

$$\Phi_n = \prod_{\mathbf{r}_i \neq 0} \left\{ \sum_j \frac{e^{-p}}{j!} \left[p\phi_n\right]^j \right\} = \prod_{\mathbf{r}_i \neq 0} \exp\left[-p + p\phi_n(\mathbf{r})\right] = \exp\left[-p \sum_{\mathbf{r}_i \neq 0} H_n(\mathbf{r})\right].$$

Using the definition of $H_n(\mathbf{r})$, Eq.(9.3), we see that $\sum_{\mathbf{r}_i \neq 0} H_n(\mathbf{r}) = \sum_{m=1}^{n} \sum_{\mathbf{r}_i \neq 0} F_n(\mathbf{r}) = \sum_{m=1}^{n} \Delta_m$ as defined by Eq.(2.20). Using now the definition of $\langle S_n \rangle$, Eq.(2.19), we get:

$$\Phi_n = \exp\left[-p\left(\langle S_n \rangle - 1\right)\right]. \tag{9.4}$$

We can use immediately the expressions for $\langle S_n \rangle$ obtained in Chapter 2, and see that in one dimension the survival probability follows a stretched-exponential pattern

$$\Phi_n = \exp\left[-p\left(\sqrt{(8/\pi)n} - 1\right)\right]$$

(see Eq.(2.22)), while in the three-dimensional case it corresponds to exponential decay:

$$\Phi_n = \exp\left[-p\left(\frac{n}{P(\mathbf{0},1)} - 1\right)\right]$$

(see Eq.(2.25)). The two-dimensional case looks almost the same: The logarithmic corrections are usually difficult to observe over a finite number of steps.

The results above can be translated into the continuous time by introducing $\phi(t,\, \mathbf{r})$ as the probability of not arriving at the donor's site up to time t, and noting that

$$\Phi(t) = \prod_{\mathbf{r}_i \neq 0} \left\{ \sum_j g(j) \left[\phi_n(\mathbf{r})\right]^j \right\} = \exp\left[-p\left(\langle S(t)\rangle - 1\right)\right] \tag{9.5}$$

where $\langle S(t)\rangle$ is the mean number of distinct sites visited by a random walker up to time t. Such a simple structure of expressions is due to the fact that different acceptors correspond to parallel channels of relaxation.

Differentiating Eq.(9.5) over time we get the differential equation for the survival probability

$$\frac{d}{dt}\Phi(t) = -p\frac{d\,\langle S(t)\rangle}{dt}\exp\left[-p\left(\langle S(t)\rangle - 1\right)\right] = -p\frac{d\,\langle S(t)\rangle}{dt}\Phi(t).$$

Multiplying the survival probability by the initial concentration of A of independent donors we get a classical form of a kinetic equation for the first-order reaction $\mathrm{A} \to 0$ catalyzed by the presence of B (note that in chemistry the catalyst is something that is not consumed, i.e., a particle whose symbol is present on both sides of the reaction equation $\mathrm{A} + \mathrm{B} \to \mathrm{B}$):

$$\frac{d}{dt}A(t) = -k(t)A(t).$$

The prefactor of $A(t)$ on the right-hand side of this equation, $k(t) = p\langle \dot{S}(t)\rangle$, is thus associated with the rate of this first-order reaction. This is actually a connection between the rates concept in chemical kinetics and the random walk observable $\langle S(t)\rangle$.

The quantity $\langle S(t)\rangle$ can be calculated by applying the methods of Chapter 3 to $\langle S_n\rangle$:

$$\langle S(t)\rangle = \sum_{n=0}^{\infty} \langle S_n\rangle\, \chi_n(t),$$

with $\chi_n(t)$ being the probability of making exactly n steps up to time t. In three dimensions, when $\langle S_n\rangle \sim n$ we thus have

$$\langle S(t)\rangle = \sum_{n=0}^{\infty} \frac{n}{P(0,1)}\chi_n(t) = \frac{\langle n(t)\rangle}{P(0,1)},$$

with $\langle n(t)\rangle$ being the mean number of steps performed up to the time t defined in Exercise 3.3, Chapter 3. For the waiting-time distributions possessing the mean waiting time τ, this is simply $\langle n(t)\rangle = t/\tau$; for the heavy-tailed power laws it is $\langle n(t)\rangle = \frac{1}{\Gamma(1+\alpha)}\frac{t^\alpha}{\tau^\alpha}$, Eq.(3.22). In the one-dimensional case $\langle S(t)\rangle$ is defined by the average of $\sqrt{n}$. Here we can use the asymptotic form $\chi_n(t) \approx \frac{t}{\alpha\tau} n^{-\frac{1}{\alpha}-1} L_\alpha\left(\frac{t}{\tau n^{1/\alpha}}\right)$ as discussed in Exercise 3.8 and the fact that

$$\langle S(t)\rangle \approx \sqrt{\frac{2}{\pi}} \int_0^\infty \frac{t}{\alpha\tau} n^{-\frac{1}{\alpha}-\frac{1}{2}} L_\alpha\left(\frac{t}{\tau n^{1/\alpha}}\right) dn = \sqrt{\frac{2}{\pi}} \frac{1}{\Gamma(1+\alpha)} \left(\frac{t}{\tau}\right)^{\frac{\alpha}{2}}$$

(Eq.(3.27)). Thus, if the acceptors perform heavy-tailed CTRW, the survival probabilities in all dimensions, also in three dimensions, are non-exponential (follow a stretched-exponential pattern).

9.3 Trapping problem

We now proceed to show that the survival probability $\Phi(t)$ depends on which of the reactants (donor or acceptor) performs the walk and which is immobile. In the target problem the donor (target) was immobile while the acceptors moved. Now we consider an inverse situation, where the donor performs the walk and the acceptors act as static traps. We specifically consider a donor that moves on a lattice, whose sites can be occupied by traps (acceptors) with probability p. The walker survives up to the n-th step of its walk if none of S_n distinct sites along its trajectory was occupied by the trap. The probability of this reads $(1-p)^{S_n-1}$ (since its initial position does not count), and the survival probability in the ensemble of walkers is given by

$$\Phi_n = \left\langle (1-p)^{S_n-1} \right\rangle .$$

This expression can be rewritten as

$$\Phi_n = \langle \exp\left[\lambda\left(S_n-1\right)\right]\rangle \tag{9.6}$$

with $\lambda = \ln(1-p)$. For $p << 1$ we have $\lambda \approx p$, so that Eq.(9.6) now reads

$$\Phi_n = \langle \exp\left[-p\left(S_n-1\right)\right]\rangle . \tag{9.7}$$

Compare this with Eq.(9.5) for the target problem: The difference is that Eq.(9.5) contains the exponential of the mean of S_n while Eq.(9.6) contains the mean value of the exponential. Therefore the inequality $\Phi_n^{\text{trapping}} > \Phi_n^{\text{target}}$ holds.[1]

Since an exponential is a (strictly) convex function, this inequality (i.e., $\langle e^x\rangle \geq e^{\langle x\rangle}$) follows immediately from Jensen's inequality, with the equality taking place only in the degenerate case of a δ-like distribution of x : $p(x) = \delta(x - \langle x\rangle)$, which is not realized in our case of fluctuating S_n.

[1]Within the Smoluchowski theory of diffusion-controlled reactions the rate of a reaction between reactants A and B depends only on the sum of their diffusivities [1], i.e., is independent of which reactant moves and which one is immobile. The discussion in this section shows that this assumption is easily violated.

The decay of A concentration in an A + B → B reaction depends on what sort of particle is mobile, and is slower in the trapping problem (A particles move, B are immobile) than in the target problem (B move, A are immobile).

We now rewrite Eq.(9.6):

$$\begin{aligned} \Phi_n &= e^{\lambda} \left\langle \exp^{-\lambda(S_n - 1)} \right\rangle \\ &= e^{\lambda} \exp\left[\sum_{j=1}^{\infty} K_{j,n} \frac{(-\lambda)^j}{j!} \right] \end{aligned} \tag{9.8}$$

where $K_{j,n}$ are the "cumulants" of the distribution of S_n defined as the Taylor coefficients of the expansion of $\ln \sum_{S_n=0}^{\infty} e^{-\lambda S_n} p(S_n)$. Truncating the expression (9.8) at the second term, we obtain for example

$$\Phi_n \approx \exp\left[-\lambda(\langle S_n \rangle - 1) + \frac{\lambda^2}{2} \sigma_n^2 \right]$$

where $\sigma_n^2 = \langle S_n^2 \rangle - \langle S_n \rangle^2$ is the variance of the number of distinct sites visited by a random walk of n steps. Note that the first term in the exponential is now the same as in Eq.(9.5) for the target problem. The positive second term corresponds to the slowing down of the reaction in the course of time. The value of σ_n^2 behaves in different dimensions as follows [2]:

$$\sigma_n^2 \propto \begin{cases} n & \text{in 1d} \\ n^2/\ln^4 n & \text{in 2d} \\ n \ln n & \text{in 3d} \end{cases} .$$

Let us now turn to the discussion of the trapping problem under CTRW. We shall assume the simplest approximation in three dimensions and neglect fluctuation effects. In this case $\Phi_n = e^p e^{-p\langle S_n \rangle}$. This is an approximation that is valid at short times, before the reaction is considerably slowed down by fluctuation effects. Then, the survival probability under CTRW reads

$$\Phi(t) \approx e^p \sum_{n=0}^{\infty} e^{-p\langle S_n \rangle} \chi_n(t) = e^p \sum_{n=0}^{\infty} e^{-pan} \chi_n(t)$$

with $a = 1/P(\mathbf{0}, 1)$ being a constant. Under Laplace transform and using the results of Chapter 3, we get

$$\begin{aligned} \Phi(s) &\approx e^p \frac{1-\psi(s)}{s} \sum_{n=0}^{\infty} \left[e^{pa} \psi(s) \right]^n \\ &= e^p \frac{1-\psi(s)}{s} \frac{1}{1 - e^{pa} \psi(s)} \end{aligned} . \tag{9.9}$$

Exercise 9.2 Show that in the case of $\psi(t)$ possessing the first moment τ the early behavior of the reaction for small p is described by $\Phi(t) \propto \exp(-pat/\tau)$.

In the case of heavy-tailed waiting-time distributions the situation changes drastically: For $\psi(s) \cong 1 - s^\alpha$, we get $\Phi(s) = \frac{e^p}{1-e^{pa}} s^{\alpha-1}$, so that

$$\Phi(t) \approx \frac{t^{-\alpha}}{pa} \tag{9.10}$$

for $p << 1$. Note that the stretched-exponential behavior in the target problem is changed for the power-law decay in the trapping problem. The behavior of the survival probability is dominated by the realizations of CTRW in which the walker stays at rest up to time t.

In the target problem in the case of heavy-tailed waiting times, the decay of A concentration follows the stretched-exponential pattern; in the trapping problem it is a power law.

9.4 Asymptotics of trapping kinetics in one dimension

The trapping problem in one dimension allows for a full solution in the continuous approximation. We consider the survival probability of a particle in a one-dimensional array of traps. The initial position of the particle is chosen at random, but wherever it is, it is confined in an interval between its two neighboring traps; all other traps play no role in our further discussion (see Fig. 9.1). The distribution of the initial particle's position within this interval is again homogeneous.

The survival probability of a particle on an interval with two absorbing boundaries was considered in Secs. 6.4 and 6.5 and is given by Eq.(6.22) in the time domain

$$\Phi(t; L) = \sum_{m=1}^{\infty} \frac{8}{\pi^2(2m+1)^2} \exp\left(-\frac{\pi^2(2m+1)^2}{L^2} Dt\right), \tag{9.11}$$

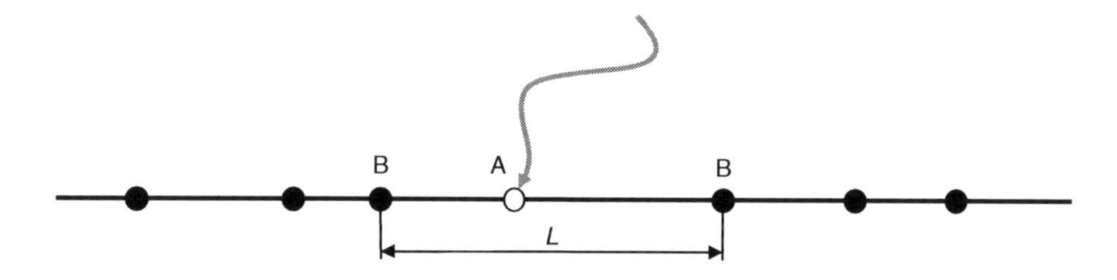

Fig. 9.1 *Trapping problem in one dimension. A newly introduced A particle is confined between the two neighboring B particles, at distance L from each other.*

or by Eq.(6.27) in the Laplace domain:

$$\Phi(s; L) = \frac{1}{s} - \frac{2\sqrt{D}}{Ls^{3/2}} \tanh\left(\sqrt{\frac{s}{D}}\frac{L}{2}\right). \tag{9.12}$$

Here we have added L to the arguments of the survival probability to indicate that it is calculated for the interval of length L.

The probability density that an A particle occurs in an interval of length L is proportional to the length of this interval and to the probability density of finding such an interval among all intervals on the line. Properly normalized this gives:

$$\pi(L) = \frac{Lp(L)}{\int_0^\infty Lp(L)dL} = \frac{Lp(L)}{\langle L \rangle}.$$

For random (Poissonian) distribution of traps with concentration p, the probability density of interval lengths is $p(L) = p\exp(-pL)$, so that $\pi(L) = p^2 L \exp(-pL)$. The survival probability averaged over all possible intervals is then given by

$$\Phi(t) = p^2 \int_0^\infty \Phi(t; L) L \exp(-pL) dL, \tag{9.13}$$

which, together with Eq.(9.11) representing $\Phi(t; L)$, is exactly the expression used in plotting Fig. 9.2. The expressions in the previous sections suggest that the behavior of $\Phi(t)$ is essentially a stretched exponential, $\Phi(t) = \exp(-f(t))$, where the function $f(t)$ is approximately a power law. This suggests the way this behavior is represented in Fig. 9.2: Plotted is $f(t) = \ln \Phi(t)$ on double logarithmic scales. The dashed and the dotted lines represent the corresponding asymptotic forms.

To assess the long-time decay behavior we note that for $t \to \infty$ only the first term in the sum in Eq.(9.11) plays a role, so that Eqs.(9.11) and (9.13) reduce to

$$\Phi(t) \approx \frac{8}{\pi^2} p^2 \int_0^\infty \exp\left(-\frac{\pi^2}{L^2} Dt - pL\right) L dL.$$

The corresponding integral can be easily evaluated by the Laplace method: The integral is of the type $\int_a^b e^{g(L)} L dL$, where the function $g(L) = -\frac{\pi^2}{L^2} Dt - pL$ has a sharp maximum at $L_0 = (2\pi)^2 (Dt/p)^{1/3}$ inside the interval of integration. Approximating $g(L)$ by the first two non-vanishing terms of its Taylor expansion, $g(L) \approx g(L_0) + \frac{1}{2} g''(L_0)(L - L_0)^2$, and expanding the interval of integration to $(-\infty, \infty)$ reduces the integral to a Gaussian one. The result of integration reads:

$$\Phi(t) \approx \frac{16}{\sqrt{3\pi}} p\sqrt{Dt} \exp\left[-\frac{3\pi^{2/3}}{2^{2/3}} p^{2/3} (Dt)^{1/3}\right]. \tag{9.14}$$

This result is shown in Fig. 9.2 as a dashed line.

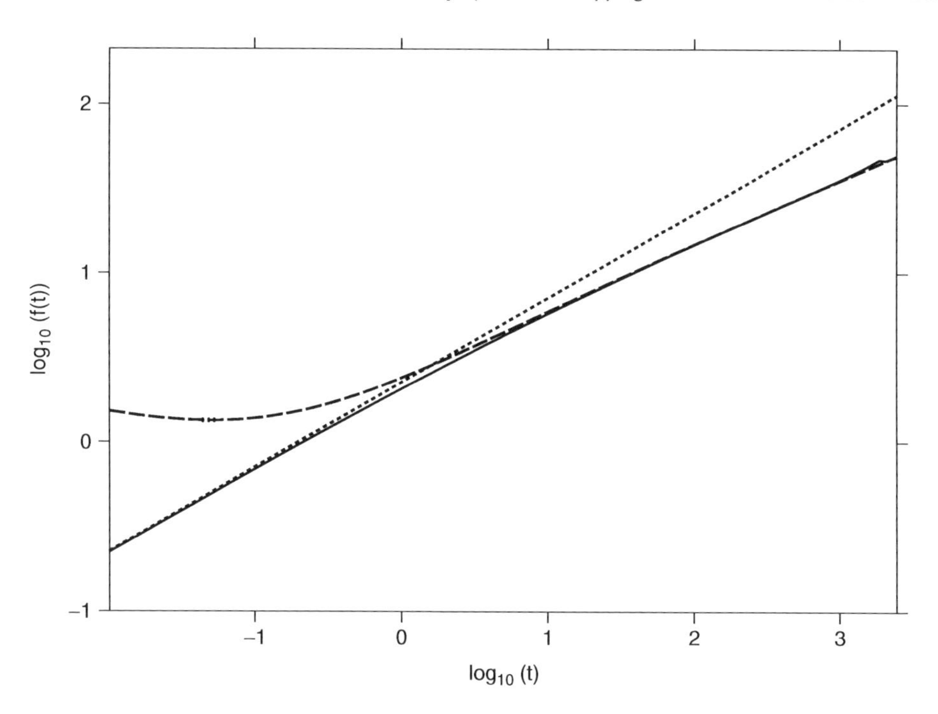

Fig. 9.2 *The behavior of the logarithm of the survival probability, $f(t) = \ln \Phi(t)$, in trapping as a function of time on the double logarithmic scale, along with its asymptotic form, Eq.(9.14), dashed line, and with the result for a target reaction, Eq.(9.15), dotted line.*

The dotted line in Fig. 9.2 shows the result that would be obtained when neglecting the fluctuation effects, i.e., that for the target problem, which for the continuum approximation would read

$$\Phi(t) \approx \exp\left[-\frac{4}{\sqrt{\pi}} p\sqrt{Dt}\right]. \tag{9.15}$$

Let us now show that for short time the survival probability $\Phi(t)$ indeed follows these asymptotics, i.e., that for short t

$$\Phi(t) \approx 1 - \frac{4}{\sqrt{\pi}} p\sqrt{Dt}.$$

It is not easy to obtain this nonanalytical dependence from a series of analytical functions such as Eq.(9.11), so another method must be used.

We start here from the Laplace representation of Eq.(9.13):

$$\Phi(s) = p^2 \int_0^\infty \left[\frac{1}{s} - \frac{2\sqrt{D}}{Ls^{3/2}} \tanh\left(\sqrt{\frac{s}{D}}\frac{L}{2}\right)\right] L \exp(-pL) dL \;, \tag{9.16}$$

where Eq.(9.12) was used for $\Phi(s;\ L)$. The behavior of $\Phi(t)$ for $t \to 0$ corresponds to that of $\Phi(s; L)$ for $s \to \infty$. In this limit, $\tanh\left(\sqrt{\frac{s}{D}}\frac{L}{2}\right) \to 1$, so that the integral over L can be taken explicitly:

$$\Phi(s) = \frac{1}{s} - \frac{2p\sqrt{D}}{s^{3/2}} .$$

The inverse transform can be easily performed using Tauberian theorem, and gives

$$\Phi(t) \cong 1 - \frac{2p\sqrt{D}}{\Gamma(3/2)} t^{1/2} = 1 - \frac{4p\sqrt{Dt}}{\sqrt{\pi}},$$

which reproduces the first term of Taylor expansion for the survival probability for the target problem in continuous limit.

The situation for subdiffusion (CTRW) can be discussed along the same lines [3], now starting from Eq.(6.29) instead of Eq.(9.10)

$$\Phi(t; L) = \sum_{m=1}^{\infty} \frac{8}{\pi^2(2m+1)^2} E_\alpha\left(-\frac{\pi^2(2m+1)^2}{L^2} K t^\alpha\right).$$

For $t \to \infty$ all Mittag-Leffler functions behave as power laws (see Eq.(6.18)), so that

$$\Phi(t; L) \cong \sum_{m=1}^{\infty} \frac{8}{\pi^2(2m+1)^2} \frac{1}{\Gamma(1-\alpha)} \frac{L^2}{\pi^2(2m+1)^2 K t^\alpha} = \frac{1}{12} \frac{1}{\Gamma(1-\alpha)} \frac{L^2}{K t^\alpha}$$

where we have used the fact that $\sum\limits_{m=0}^{\infty} \frac{1}{(2m+1)^4} = \frac{\pi^4}{96}$ (Eq.(0.234.5) of Ref.[4]). Averaging this form over the distribution of L, Eq.(9.13), we obtain

$$\Phi(t) = \frac{1}{2\Gamma(1-\alpha)} \frac{1}{p^2 K t^\alpha}.$$

This is a one-dimensional analog of our Eq.(9.10). The different dependence on p is due to the recurrence of walks in one dimension.

9.5 Trapping in higher dimensions

In higher dimensions only approximate results on the asymptotics of the survival probability in a trapping problem are known. Contrary to the one-dimensional case, where the problem is reduced to the exactly solvable problem of survival in the interval with absorbing boundaries, in higher dimensions no such simplification is possible. However, we can still use the same method to obtain *the lower bound* for this survival probability.

Let us consider the initial position of the A particle and its surroundings, and let us consider the largest d-dimensional ball (a circle in two dimensions, a ball in three dimensions, etc.) centered on A and that contains no B particles. Let L be the radius of such a ball. As long as the particle performs walks within this ball, it definitely survives. If the displacement of the particle from its initial position reaches L it may (but does not have to) react. Therefore considering the sphere at L as an absorbing boundary, we can get a lower bound on the survival probability of the particle. The discussion here follows that in Sec. 9.4 and the eigenfunction decomposition discussed in Chapter 6.

The survival probability of a particle within a ball is given by a diffusion equation

$$\frac{\partial p(\mathbf{r},t)}{\partial t} = K\Delta p(\mathbf{r},t) \tag{9.17}$$

with the initial condition $p(\mathbf{r},0) = \delta(\mathbf{r})$ and the boundary condition $p(L,t) = 0$. Although the corresponding boundary value problem can be solved explicitly (in special functions) in $d = 2$ and $d = 3$, we refrain from this, because of the roughness of our initial approximation.

The long-time decay of the overall concentration is given by the exponential form

$$\Phi(t) \propto \exp(-|\lambda_0(L)|\, t)$$

where $\lambda_0(L)$ is the lowest eigenvalue of the spatial part of the corresponding equation:

$$K\Delta X_n(\mathbf{r}) = \lambda_n(L)X_n(\mathbf{r})$$

with the boundary condition $X_n(\mathbf{r})|_{|\mathbf{r}|=L} = 0$. First, from the form of this equation it follows that the eigenvalues are proportional to the diffusion coefficient K. Imagine that the corresponding eigenvalue $\lambda_0(1) = const \cdot K$ and eigenfunction $X_0(\mathbf{r})$ are known for $L = 1$. Then those for any other L follow by the change of variables, namely passing to a new variable $\boldsymbol{\rho} = \mathbf{r}/L$ that reduces the new equation to the previous one, for $L = 1$. The eigenvalue of the equation in the initial units is thus $\lambda_0(L) = const \cdot K/L^2$, which leads to an exponential form

$$\Phi(t;L) \propto \exp\left(-const\,\frac{Kt}{L^2}\right). \tag{9.18}$$

This has then to be averaged over the distribution of the sizes of such particle-free balls. The last one is given by the ball's volume V (circle's surface in two dimensions): $p(V) = p\exp(-pV)$, where p is the B-particle concentration. Noting that the volume of the d-dimensional ball is connected to its radius via $V = \Omega(d)L^d$, with $\Omega(d)$ being a prefactor depending on the dimension of space ($\Omega(d) = \pi$ in $d = 2$; $\Omega(d) = 4\pi/3$ in $d = 3$), we get

$$p(L) = p^2 d\Omega(d)L^{d-1}\exp(-p\Omega(d)L^d)\ .$$

Now evaluating

$$\Phi(t) = p^2\int_0^\infty \Phi(t;L)d\Omega(d)L^{d-1}\exp(-p\Omega(d)L^d)\,dL\ ,$$

we obtain (up to pre-exponential)

$$\Phi(t) \propto \exp\left[-C(d)p^{\frac{2}{d+2}}K^{\frac{d}{d+2}}t^{\frac{d}{d+2}}\right] \tag{9.19}$$

with $C(d)$ being a constant depending only on the dimension of space.

Exercise 9.3 Use the Laplace method outlined in Sec. 9.4 to prove Eq.(9.19).

References

[1] S.A. Rice. *Diffusion-Limited Reactions*, Amsterdam: Elsevier, 1985

[2] B.D. Hughes. *Random Walks and Random Environments*, Vol. 1: *Random Walks*, Oxford: Clarendon, 1996 (The information discussed is in Sec. 6.2 of this book.)

[3] A. Blumen, J. Klafter, and G. Zumofen. "Reactions in Disordered Media Modelled by Fractals," in: Pietronero, L., and Tosatti, E., eds., *Fractals in Physics*, Amsterdam: North-Holland, 1986, pp. 399–408

[4] I.S. Gradstein and I.M. Ryzhik. *Table of Integrals, Series and Products*, Boston: Academic Press, 1994

Further reading

G.H. Weiss. *Aspects and Applications of the Random Walk*, Amsterdam: North-Holland, 1994

W. Ebeling and I.M. Sokolov. *Statistical Thermodynamics and Stochastic Theory of Nonequilibrium Systems*, Singapore: World Scientific, 2005 (Relevant material is in Chapter 10 of this book.)

10 Random walks on percolation structures

"The question is too good to spoil it with an answer."

Robert Koch

Up to now we have discussed random walks in homogeneous spaces. Let us now consider a simple random walk on a lattice and start diluting the lattice. By this we mean that some fraction of the lattice sites is removed, i.e., is declared inaccessible for the walker. As long as the concentration of forbidden sites is low, nothing spectacular happens: For example, the MSD of the walker still grows as

$$\langle \mathbf{r}^2(n) \rangle \cong C(p)n \tag{10.1}$$

with the prefactor $C(p)$ now depending on the concentration of sites still accessible for the motion. We can also translate this behavior to continuous time. For example, if the mean waiting time τ for the step exists (see Chapter 3), the relation

$$\langle \mathbf{r}^2(t) \rangle \cong dK(p)t \tag{10.2}$$

holds. Here $K(p) = C(p)/d\tau$ is the effective diffusion coefficient, depending on the concentration of the accessible sites (d is the dimension of space). There exist quite effective approximations that allow for evaluating $K(p)$ for p close enough to unity [1]; however, this is not our topic here.

Let us continue removing sites. At some concentration p the lattice disintegrates, and the diffusion over large distances becomes impossible. Indeed, for small p only small finite clusters of accessible sites (single sites, dimers, triples, etc.) are present, all surrounded by "firewalls" of inaccessible ones, leaving no possibility for unbounded motion. The cluster is defined as a set of connected accessible sites, and is essentially the part of the lattice that a walker placed on one of these sites can explore. The concentration $p = p_c$ at which the infinite cluster appears when p grows (i.e., at which the unbounded motion through the lattice becomes possible) is called the percolation concentration, or the percolation threshold. The percolation threshold p_c is a well-defined quantity for each lattice, and is known for many simple lattices (see Table 10.1).

Some of these concentrations are known exactly, other ones as results of numerical simulations. The investigation of percolation transition is the subject of *percolation theory* [2] that has now grown into a well-developed tool for theoretical investigation

Table 10.1 Percolation concentrations for some common lattices

Lattice	p_c **in site percolation**	p_c **in bond percolation**
triangular	$1/2$	$2\sin(\pi/18) \approx 0.347$
square	0.593	$1/2$
honeycomb	0.697	$1-2\sin(\pi/18) \approx 0.653$
diamond	0.43	0.39
simple cubic	0.312	0.249
bcc	0.246	0.180
fcc	0.199	0.120

of structural disorder. Parallel to the *site* problem discussed above, the *bond* problem of percolation can be formulated, in which all sites are present, but some fraction of bonds, connecting them, is removed, thus leading to disintegration of the lattice.

A physical realization of random walks in a percolation system corresponds, e.g., to diffusion of optical excitations in mixed molecular crystals (binary substitutional disorder [3]), where the excitation A can be present on molecules of only one type (accessible sites) and cannot be transferred to those of the other type [4]. Since excitations can also react, this realization later serves as an example of reaction on percolation clusters. Let us first concentrate on diffusion without reaction.

Let us imagine that a walker starts on an infinite cluster and travels on it. The prefactor $C(p)$ in Eq.(10.1) or the diffusion coefficient $K(p)$ in Eq.(10.2) decreases with decreasing p and vanishes exactly at the percolation threshold $p = p_c$, when the lattice disintegrates. Exactly at this concentration, the transport is still possible but corresponds to another type of n-dependence:

$$\langle \mathbf{r}^2(n) \rangle \propto n^\alpha \tag{10.3}$$

with $\alpha < 1$. Translating this into time leads to $\langle \mathbf{r}^2(t) \rangle \propto t^\alpha$, a subdiffusive behavior. This is yet another type of subdiffusion compared to that encountered in CTRWs with broad distribution of waiting times, as relates, e.g., to energetic disorder (Chapter 3).[1] The situation would be quite uninteresting if Eq.(10.3) related to a special, single point, the percolation threshold. This is, however, not the case. For p close enough to p_c the MSD follows Eq.(10.3) for quite a while before it crosses over to normal diffusion Eq.(10.1) on an infinite cluster above p_c or stagnates on a large but finite cluster below p_c.

[1] Trying to qualitatively explain this behavior, one often refers to a simple picture representing the infinite cluster at the threshold as consisting of a backbone (being a loose and tortuous structure through which infinite motion is possible) and dangling ends, connected to the backbone at only one site. The subdiffusion then is considered to be due to the tortuosity of the backbone, as well as to trapping in the dangling ends, making the system look a bit like the comb considered in Chapter 3. The analogy with the comb is, however, misleading, since it would lead to aging phenomena (ergodicity breaking), which are not observed on percolation clusters (see Ref.[5] for a detailed discussion of the problem). Moreover, motion on a Sierpinski gasket is also subdiffusive (see Sec. 10.2), although in this case no dangling ends exist.

The theoretical description in this chapter does not follow the rigorous methods outlined in the previous chapters (therefore the motto of this one). Most of the facts discussed here are *known* from numerical simulations and *believed* to be universally valid. These are only a few of those proved at the mathematical level of rigor [1]. For a physicist, however, it is very important that a special language and tools have been developed for reasoning in such situations, those based on concepts of fractal geometry. These are very useful when coping with situations pertinent to structural disorder.

10.1 Some facts about percolation

In this section we proceed with a discussion of some interesting properties of percolation structures close to the percolation threshold. Our discussion is by no extent full; the interested reader is sent to [1, 2]. Looking at the percolation system at $p > p_c$, we see that it consists of the infinite cluster (a large connected part taking the finite portion of the overall volume) with holes in which the smaller clusters sit. The infinite cluster with holes (but without finite clusters) is shown in the right-hand panel of Fig. 10.1.

It is proved that under typical conditions there is a unique infinite cluster in a system, if any. The typical size of the holes defines the characteristic scale in a system, which we will call the correlation length ξ. When removing all finite clusters that do not contribute to the unbounded motion, and looking at the structure of the infinite one at scales larger than ξ, this will appear homogeneous, with the density $p_\infty < p$. The density p_∞ of the infinite cluster is nothing other than the probability of picking up a site belonging to it. Below the percolation threshold the infinite cluster does not exist and $p_\infty = 0$. Close to the percolation threshold the density p_∞ behaves as a power law

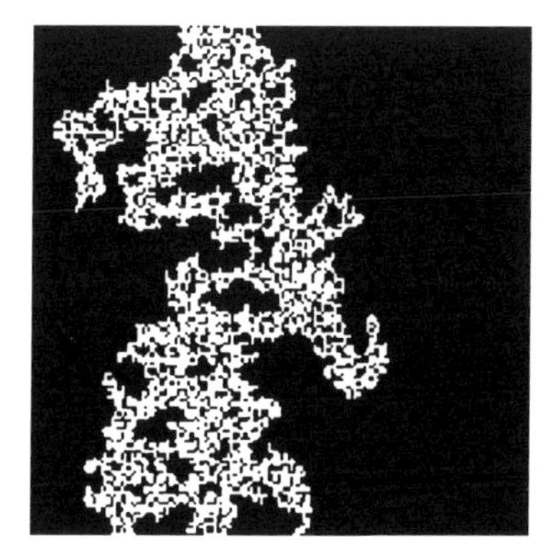

Fig. 10.1 *The structure of percolation system close to p_c: The left-hand panel shows all clusters (white sites) on a 150×150 square lattice, the right-hand panel only the largest ("infinite") cluster separated from the rest. Note that this has holes on all scales (is fractal).*

Table 10.2 Commonest critical exponents of percolation theory

Exponent	$d = 2$	$d = 3$
β	5/36	0.41
γ	43/18	1.80
ν	4/3	0.88
τ	187/91	2.186

$$P_\infty \propto (p - p_c)^\beta, \tag{10.4}$$

where the values of critical exponent β are known (at least numerically) in all relevant dimensions (see Table 10.2). While approaching the critical point from above, the density of the percolation cluster decays, which means that the holes in it get larger, so that ξ diverges. This divergence also follows a power law, $\xi \propto (p - p_c)^{-\nu}$.

In fact the standard definition of the correlation length is slightly different: ξ can be defined as the mean gyration radius of a finite cluster, and thus is finite on both sides of the percolation transition:

$$\xi \propto |p - p_c|^\nu . \tag{10.5}$$

Since above the percolation transition the holes in the infinite cluster are filled by finite ones, both definitions coincide. The properties of a percolation system close to the transition (from both sides) are defined by the properties of large finite clusters, discussed below.

Close to the percolation transition the probability of finding a finite cluster with N sites among all finite clusters scales as

$$p_N \propto N^{-\tau} f(-G(p)N) \tag{10.6}$$

where τ defines a new critical exponent. The function f decays fast enough for large negative values of its argument and describes a characteristic cutoff in the power-law distribution $p_N \propto N^{-\tau}$ which holds exactly at p_c. The prefactor $G(p)$ characterizes the inverse number of sites in the largest finite clusters and vanishes at the percolation threshold. $G(p)$ shows a power-law dependence on the concentration:

$$G(p) \propto |p - p_c|^{1/\sigma} .$$

The exponent σ and especially τ, which is repeatedly used in what follows, are connected with the more common ones defining the behavior of the moments of cluster size distribution, which are often given in the literature. For example, the exponent γ defines the divergence of the second moment of the cluster size:

$$S(p) = \frac{1}{p_c} \sum_{N=1}^{\infty} N^2 p_N \propto |p - p_c|^{-\gamma} \tag{10.7}$$

(the infinite cluster is excluded from summation). Thus,

$$\sigma = \frac{1}{\beta + \gamma}$$

and

$$\tau = 2 + \frac{\beta}{\beta + \gamma} \tag{10.8}$$

(see Ref.[2]).

The values of the corresponding exponents in two and three dimensions are taken from Ref.[2] and given in Table 10.2.

10.2 Fractals

Above percolation concentration, the infinite cluster is characterized by finite density, and the number of sites ("mass") of the part of the cluster inside the ball of radius L grows with this radius as

$$M = const(d) \cdot p_{\infty} L^d \tag{10.9}$$

where $const(d)$ depends only on the dimensionality of space. The dependence of the mass on the size of the system is characteristic of the spatial dimension, which can be obtained, e.g., in an experiment by measuring M at different sizes L. Thus, increasing L by a factor of 2 increases M by a factor 2^d, and

$$d = \frac{\ln(M(L_1)/M(L_2))}{\ln(L_1/L_2)}. \tag{10.10}$$

We can argue that, instead of cutting parts of different sizes from the infinite system, we can add parts together and define the dimension as an exponent describing the growth of the mass of a finite system put together from smaller but similar parts. For example, a square can be put together from four similar squares of half the size, and therefore its dimension is $d = \ln 4/\ln 2 = 2$, while the dimension of a cube, which can be put together from eight cubes of half the size again, is $d = \ln 8/\ln 2 = 3$. The dimension defined through the mass of the object (or through the number of similar parts that it is built of) is called the mass dimension.

The simple example in Fig. 10.2 depicts an object (called the *Sierpinski gasket*) constructed from three similar objects of half the size and therefore having a dimension $d = \ln 3/\ln 2 \approx 1.585\ldots$, which is not a whole number. The objects with non-integer (fractional) mass dimensions are called fractals. Mass dimension is one of the family of differently defined fractal dimensions, and is considered as *the* fractal dimension in what follows.[2]

[2]Mathematicians introduce dimensions and related properties by covering procedures, e.g., by counting how the number of boxes, or balls, etc., covering the system will grow when the box's or ball's size tends to zero. There are mathematical reasons to do so (e.g., to let the dimension be invariant under smooth coordinate transformations), but no physical grounds to follow this reasoning. All physical systems consist of atoms, and these are not too small.

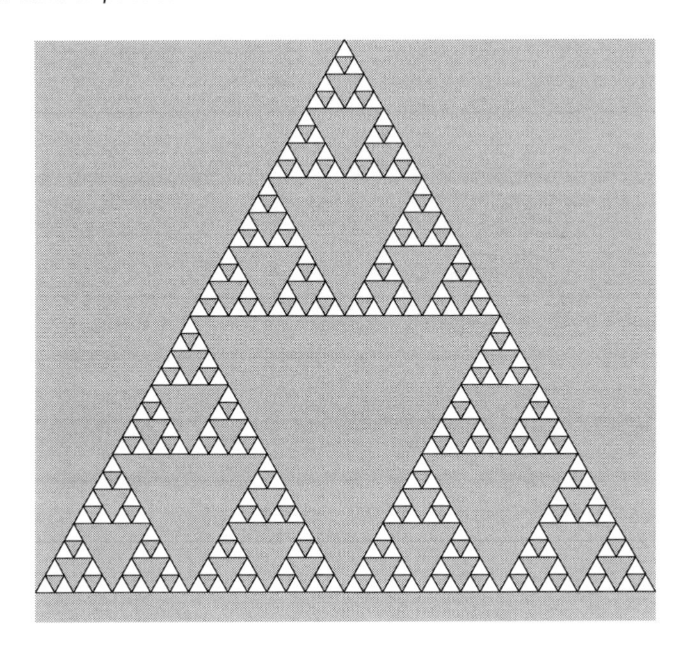

Fig. 10.2 *A Sierpinski gasket. An object of n-th generation consists of three objects of $(n-1)$-th generation put together. The objects of the first generation are the white triangles.*

Exercise 10.1 The structures analogous to the Sierpinski gasket can be defined also in higher dimensions, e.g., in $d = 3$ when it has the overall form of a tetrahedron built of smaller ones, and even in dimensions higher than three. Calculate the fractal dimension of the Sierpinski construction in d dimensions.

Hint: Consider a recurrence relation between different d.

The fractal dimension of the infinite percolation cluster can be obtained by the following argument (well supported by numerical data). Let us consider the cluster at p slightly above p_c, when its density is finite but small. The structure of the cluster will reproduce that of the cluster at p_c at small scales (in small volumes the concentration of the accessible sites shows statistical fluctuations, and the cluster simply does not "know" that it is not exactly at p_c), while at larger scales this density will stagnate at the value of p_∞. To grasp the properties of such a structure imagine a plane covered with tiles in the form of the Sierpinski gasket with the side L. This system is fractal on small scales and becomes homogeneous on scales of the order of L.

In the fractal regime the mass of a piece of an infinite cluster of size L grows with L as $M(L) \propto L^{d_f}$, and thus its density behaves as

$$\rho(L) \propto L^{d_f - d}$$

until L becomes of the order of ξ. At larger scales the cluster is homogeneous with density p_∞. Thus, $p_\infty \sim \xi^{d_f - d}$. Now we can express p_∞ and ξ as functions of $p - p_c$ and get $(p-p_c)^\beta \sim (p-p_c)^{\nu(d_f - d)}$ so that

$$d_f = d - \frac{\beta}{\nu}. \tag{10.11}$$

This dimension is equal to 91/48 in $d = 2$ and around 2.53 in $d = 3$. In dimensions higher than 6 the value of d_f stagnates at $d_f = 4$; this is, however, not a situation we consider here.

10.3 Random walks on fractals

Random walks on fractals have been extensively investigated in hundreds of works, on different structures. The results of the simulations can be summarized as follows.

The MSD of a random walker on a fractal grows subdiffusively (Eq.(10.3)). This equation is often rewritten in a different way. Let us consider the mass dimension of the random walker's trajectory by inverting Eq.(10.3). A piece of trajectory of n steps (i.e., that of "mass" n) has a typical spatial dimension $L \sim n^{\alpha/2}$. Thus, the mass of the piece of a trajectory of size L will go as $n \propto L^{2/\alpha}$ defining the fractal dimension of the walker's trajectory (*walk dimension*) $d_w \propto 2/\alpha$. Therefore Eq.(10.3) can be rewritten as

$$\left\langle \mathbf{r}^2(n) \right\rangle \propto n^{2/d_w}. \tag{10.12}$$

In subdiffusive motion ($\alpha < 1$) typical of fractals, the walk dimension always exceeds 2.[3]

In the case of random walks on regular lattices, the probability of finding the walker at the origin behaves as $P_n(0) \propto \frac{1}{(\sqrt{n})^d} = n^{-d/2}$, i.e., follows the power-law decay governed by the spatial dimension of the system. In a fractal system the power-law decay is also observed, and the long-time asymptotics of this decay

$$P_n(\mathbf{0}) \cong A n^{-d_s/2} \tag{10.13}$$

(with A being a constant) defines a new characteristic of a fractal system, its *spectral dimension*.[4] In the case of Euclidean lattices, $d_f = d_s = d$, but for fractals, d_f and

[3]The subdiffusion as described by CTRW models arises from waiting times, during which the walker does not move. The trajectory of the walker is exactly that of the simple random walk, and as a function of n we have $\left\langle \mathbf{r}^2(n) \right\rangle \propto n$. Thus, the walk dimension of CTRW is 2. This gives us a method to distinguish between the types of disorder pertinent to CTRW and to diffusion on fractal substrates. Interestingly enough, the relation $\left\langle \mathbf{r}^2(n) \right\rangle \propto n$ is also true for the comb, independently of whether steps on a spine (CTRW) or the steps in the teeth are also taken into account: Here only the prefactors differ. Once again: A comb is an interesting model, but not the relevant one for percolation!

[4]The term "spectral dimension" refers to the fact that the same quantity governs the density of low-frequency modes on a fractal considered as an elastic structure [6], and has to do with the spectral properties of discrete Laplace operator on the structure (i.e., it is a purely geometrical quantity). Modern mathematical works, however, define d_s exactly as given in Eq.(10.13).

d_s are somewhat independent characteristics, although they are connected by an inequality

$$d_s \leq d_f \tag{10.14}$$

(note that $d_f > 1$ for any structure on which an unbounded random walk is possible). The bound follows from the fact that $d_w \geq 2$. For percolation clusters in any dimension, d_s is very close to 4/3. Although the initial assumption that it is exactly 4/3 [6] has been proven to be incorrect, with the largest deviation of around 2% observed for $d = 2$ [7], the Alexander–Orbach conjecture $d_s = 4/3$ can still be considered exact for practical purposes.

Numerical evidence shows that the PDF of the distance from the origin scales. This means that the probability of finding the walker at distance r from the initial site is given (at least asymptotically) by the following form:[5]

$$P_n(r) = An^{-d_s/2} f\left(\frac{r^2}{\langle r^2(n)\rangle}\right) = An^{-d_s/2} f\left(\frac{r^2}{n^{2/d_w}}\right), \tag{10.15}$$

where the function $f(x)$ can be chosen in such a way that $f(0) = 1$ to comply with Eq.(10.13). Equation (10.15) allows us to find the connection between d_s and d_w. Indeed, let us assume that the function $f(x)$ decays fast enough for $x \to \infty$ (a Gaussian decay in the diffusion on regular lattices being an example). Then most sites visited by the random walker are situated within a ball with a radius of around $L \cong n^{1/d_w}$, and the overall number of such sites is given by the fractal dimension of the system: $N \cong L^{d_f} \cong n^{d_f/d_w}$.

The two main geometric characteristics of a fractal lattice, its fractal dimension d_f and its spectral dimension d_s, define the properties of the random walk on this structure. The MSD of the walker behaves as $\langle \mathbf{r}^2(n)\rangle \propto n^{2/d_w}$. Here d_w is the fractal dimension of the walker's trajectory and is given by $d_w = \frac{2d_f}{d_s}$.

The spectral dimension of a percolation cluster is close to 4/3.

Assuming that all these sites can be visited with a comparable probability and that the origin is just one of these more or less equivalent sites, we conclude that $P_n(0) \propto 1/N \cong n^{-d_f/d_w}$. Comparing this with Eq.(10.13) we get

$$d_w = \frac{2d_f}{d_s} \tag{10.16}$$

[5]Much effort was put into deriving the equation for this function f in the continuous limit, i.e., the corresponding generalized diffusion equation. The first and the simplest one was proposed in Ref.[8]. Newer simulations, however, show that the function f is a property of the structure and indicate that there may be no universal equation for it. Some authors have proposed equations containing time fractional derivatives, of the type considered in Chapter 6. These need some reservation: The process of random walks on a fractal considered here shows no aging effects typical of CTRW processes [5] for which such equations are adequate.

connecting the properties of walks with the geometric properties of the fractal substrate.

10.4 Calculating the spectral dimension

Evaluation of the probability of being at the origin is not the best way to evaluate the spectral dimension numerically. Since argumentation in terms of continuous time (leading to a customary master equation as a description tool) is easier to follow, we adopt this here. Asymptotically, the t- and the n-dependencies are of course the same.

The master equation for the probabilities p_i of finding a particle at a site i reads

$$\frac{d}{dt}\mathbf{p} = \mathbf{W}\mathbf{p}, \tag{10.17}$$

where $\mathbf{p}$ is the vector comprising all these p_i. The matrix $\mathbf{W}$ describes the transition probabilities between the nodes of the network. The non-diagonal elements of the corresponding matrix give the transition probabilities from site i to site j per unit time, and are assumed to be unity for connected sites and zero otherwise. The diagonal elements are the sums of all non-diagonal elements in the corresponding column taken with the opposite sign:

$$w_{ii} = -\sum_{j \neq i} w_{ij}, \tag{10.18}$$

which represents the probability conservation law. The same equation (10.17) of course holds for the mean numbers of particles on the site if a many-particle situation is considered; we have only to change the probability p_i of finding a particle on the site i for the mean number of particles n_i on the site, which will comprise a vector $\mathbf{n}$.

Up to now, to define the spectral or the walk dimension we have used the time-dependent setup and looked at the temporal evolution of the particle's MSD, of the mean number of sites visited, etc. We can try, however, to revert the equation for the root MSD $L \propto t^{1/d_w}$, and calculate d_w via the mean time T necessary to travel between a given point and an absorbing boundary of a finite system at a distance L from the origin, and hope that in non-pathological cases $T \propto L^{d_w}$. Moreover, instead of following particles one by one, we can introduce them, say, one per unit time, i.e., consider the constant particles' current I. After some time, a stationary distribution of particles will establish in the system. This is described by a stationary version of Eq.(10.17)

$$\mathbf{W}\mathbf{n} = 0 \tag{10.19}$$

with the boundary conditions corresponding to particles' introduction at the origin and particles' absorption at an outer boundary. The overall number of particles in the system is $N = \sum_i n_i$ and defines the mean lifetime of the particles T via $N = IT$, so that T can be resolved as $T = N/I$.

Calculating concentrations n_i formally corresponds to calculating the voltages on the nodes of the resistor network of the same geometry under a given overall current using Kirchhoff's laws. The fact that the probability current between nodes i and j is proportional to the difference $p_j - p_i$ is replaced by Ohm's law. The conductivities of resistors connecting nodes i and j have to be taken to be 1 if the nodes are connected. The condition given by Eq.(10.19) together with the condition in Eq.(10.18) corresponds then to Kirchhoff's second law representing particle conservation (the fact that the sum of all currents to/from the node i is zero), and Kirchhoff's first law follows from the uniqueness of the solution. Therefore the dimension of a walk can be calculated by discussing the scaling of resistivity R of the fractal object with its size L. This typically follows a power law:

$$R \propto L^{\zeta}. \tag{10.20}$$

To connect the conductivity exponent ζ with the walk dimension, we can make the following reasoning. Let us assume that the distribution of the particles' concentration in fractals of different size is similar. The mean number of particles N inside the system is then proportional to a typical concentration (which in its turn will be proportional, say, to the concentration n_A of the particles at the current injection point) and to the number of sites. The first one, for a given current, scales as the system's resistivity, $n_A \sim IR \propto L^{\zeta}$. On the other hand, the number of sites (the mass of the fractal) scales as $M \propto L^{d_f}$. Thus $T \sim N \propto L^{\zeta + d_f}$, so that the walk dimension is given by

$$d_w = \zeta + d_f. \tag{10.21}$$

As an example let us calculate d_s of a Sierpinski gasket. To do this we first have to determine ζ, i.e., to calculate the resistance between the terminals of a fractal of the next generation (depicted by thick wires outside of the triangle in Fig. 10.3), assuming that the resistance between the corresponding nodes of the lattice of the previous generation is known (let us take this to be r). This can be done using a triangle–star transformation, an old trick known in the theory of electric circuits. We simply pass to the structure shown in Fig. 10.3 by thick lines inside the triangle, with the conductivity of each bond being $r/2$, which gives us the same value r of the resistance between the terminals. The resistance of a structure of the next generation is then easily calculated and equals $\frac{5}{3}r$. Thus, the dependence of R on the spatial scale L of the object is $R \propto L^{\zeta}$ with $\zeta = \log(5/3)/\log 2$. Since the fractal dimension of the gasket is $d_f = \log 3/\log 2$, the walk dimension is $d_w = \log 5/\log 2$ and the spectral one equals $d_s = 2\log 3/\log 5$.

Exercise 10.2 Consider d-dimensional generalizations of Sierpinski gaskets (Exercise 10.1). Show that for them

$$d_w = \frac{\log(d+3)}{\log 2}.$$

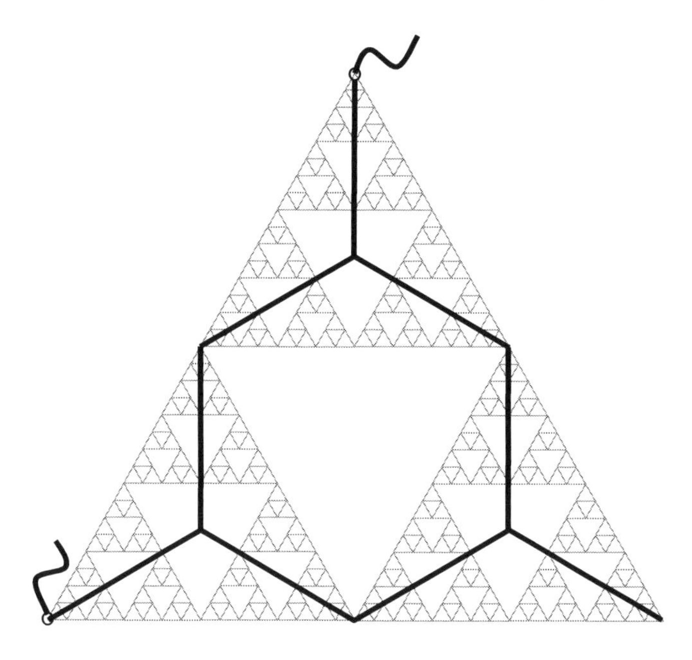

Fig. 10.3 *Triangle–star transformation used in a calculation for scaling the conductivity of the Sierpinski gasket with its size.*

Spectral dimensions of many other structures are obtained by similar methods, for example, those of percolation clusters by numerical methods evaluating the resistance of complex networks. In two dimensions especially effective algorithms are known [9].

The simplest way to calculate the walk dimension and the spectral dimension is by considering the fractal as a conducting structure and determining how its resistance R scales with its size L. If $R \propto L^{\zeta}$ then $d_w = \zeta + d_f$ and $d_s = \frac{2d_f}{d_f+\zeta}$.

10.5 Using the spectral dimension

Discussion of Eq.(10.16) leads to another interesting property regarding spectral dimension. Let us consider the number of sites $N \propto \left(n^{1/d_w}\right)^{d_f} = n^{d_s/2}$ within the ball with the radius $\left\langle \mathbf{r}^2(n) \right\rangle^{1/2}$ for the walk of n steps. For $d_s < 2$ the number of sites grows more slowly than the number of steps. Consequently, virtually all sites within the ball are visited (a property of *compact visitation*), and each of them is visited several times (*recurrence* property). Both of these properties are known to us from the results for one-dimensional systems. In contrast, for $d_s > 2$ the number of sites within the ball grows asymptotically faster than the number of steps. Hence, not all of them can be visited by the random walk, and those that are visited can only be revisited a finite number of times. Thus, the number of distinct sites visited behaves as

$$S_n \propto \begin{cases} n^{d_s/2} & \text{for } d_s < 2 \\ n & \text{for } d_s > 2 \end{cases}, \tag{10.22}$$

as is known for random walks on regular lattices (with d_s standing for spatial dimension d). Equation (10.22) allows for qualitative discussion of all situations in which S_n plays a role, for example that of reactions.

Random walks are recurrent on fractals with $d_s < 2$ and transient on fractals with $d_s > 2$.

Recurrent random walks on fractals have a lot in common with the known situation of recurrent walks in one dimension. Thus, the probability distribution of returning to the origin can be obtained using the renewal approach, relying on the Markov character of such walks. On random fractals such as percolation clusters all quantities have to be considered as averages over the initial positions.

Writing down the renewal equation in discrete time (number of steps)

$$P_n(\mathbf{0}) = \delta_{n,0} + \sum_{k=1}^{n} F_k(\mathbf{0}) P_{n-k}(\mathbf{0}),$$

(cf. Eq.(2.8)) and moving on to a generating function lets us express the properties of the first return distribution by spectral dimension: Since Eq.(2.11) holds,

$$F(\mathbf{0}, z) = 1 - \frac{1}{P(\mathbf{0}, z)},$$

and since for $P_n(\mathbf{0}) \cong An^{-d_s/2}$, Eq.(10.13), the Tauberian theorem gives $P(\mathbf{0}, z) \cong A\Gamma(1 - d_s/2)(1 - z)^{d_s/2-1}$, we get $F(\mathbf{0}, z) = 1 - \frac{1}{A\Gamma(1-d_s/2)}(1 - z)^{1-d_s/2}$, which in its turn corresponds to

$$F_n(\mathbf{0}) \cong -\frac{1}{A\Gamma(1 - d_s/2)\Gamma(d_s/2 - 1)} n^{d_s/2-2}$$

(note that $\Gamma(d_s/2 - 1) < 0$ for $d_s < 2$). This is the generalization of Sparre Andersen's result (2.35) to fractal lattices with $d_s < 2$.

The fact that d_s governs the mean number of distinct sites visited allows us to extend the results of Chapter 9 to fractal lattices as well. For a target problem, where the survival probability $\Phi_n = \exp\left[-p\left(\langle S_n \rangle - 1\right)\right]$ is exactly governed by the mean number of distinct sites, this translation is trivial, so that

$$\Phi_n \propto \begin{cases} \exp\left(-const \cdot pn^{d_s/2}\right) & \text{for } d_s < 2 \\ \exp\left(-const \cdot pn\right) & \text{for } d_s > 2 \end{cases}.$$

The translation to the continuous time (if necessary) follows in exactly the same way as for walks on regular lattices considered in Chapter 9.

In the case of trapping reactions, we have to repeat the lines of argumentation used in Section 9.5. Thus, the mean time τ (or the mean number of steps ν) a walker survives in a trap-free part of a fractal within the ball of radius L scales as $\tau \propto L^{d_w}$. The form of the asymptotical survival probability

$$\Phi(t;L) \propto \exp\left(-const\,\frac{t}{L^{d_w}}\right)$$

at longer times is still exponential. This can be assumed due to the fact that at times of the order of τ the particle explores practically the whole volume available, and the distribution of its positions (provided it survived) tends to some stationary form. The survival probabilities over two subsequent τ-intervals are the same, and the survival probability over many such intervals is a product thereof, i.e., tends to the exponential form in time.

Noting that the volume (number of sites) of the d_f-dimensional compartment of size L is connected to L via $V \propto L^{d_f}$, we obtain

$$p(L) = const \cdot d_f p L^{d_f - 1} \exp(-const \cdot p L^{d_f}) \; .$$

Applying the Laplace method for calculation of the corresponding integral, we get (up to a pre-exponential)

$$\Phi(t) \propto \exp\left[-const \cdot p^{\frac{d_w}{d_f+d_w}} t^{\frac{d_f}{d_f+d_w}}\right] \; ,$$

which can be rewritten using the relation (10.16):

$$\Phi(t) \propto \exp\left[-const \cdot p^{\frac{2}{d_s+2}} t^{\frac{d_s}{d_s+2}}\right] \; .$$

This is exactly Eq.(9.19) with the spatial dimension d exchanged for d_s.

10.6 The role of finite clusters

An interesting property that makes a percolation system close to p_c different from regular fractals is connected with the role of finite clusters. Depending on the physical problem at hand, we can consider situations where only the infinite cluster plays the role (when for example only the infinite cluster in a porous system is filled by fluid pressed through the boundary of the system and all interesting reactions take place in this fluid), and situations where the excitations can be found in infinite as well as in finite clusters, as in the case of optical excitations (which are generated both in infinite and in finite clusters) in mixed molecular crystals. Here we consider only the situation at $p = p_c$ exactly.

The structure of large but finite clusters is similar to that of the infinite one, so that, for example, the MSD on the finite cluster grows as

$$\left\langle \mathbf{r}^2 \right\rangle \sim n^{1/d_w}$$

as long as $\left\langle \mathbf{r}^2 \right\rangle^{1/2}$ is smaller than the cluster's typical size L, and then stagnates at the value $\left\langle \mathbf{r}^2 \right\rangle \sim L^2$ when this boundary is reached, i.e., for $n \sim L^{d_w}$. Using the number of sites N in a cluster as a measure of its size instead of $L \sim N^{1/d_f}$, we get

$$\left\langle \mathbf{r}^2 \right\rangle \sim \begin{cases} n^{1/d_w} \text{ for } n < N^{2/d_s} \\ N^{1/d_f} \text{ for } n > N^{2/d_s} \end{cases}. \tag{10.23}$$

The probability that a particle starts on a cluster of N sites is proportional to $Np(N)$, with $p(N)$ being the probability of finding a cluster of N sites among all clusters. This one goes as $p(N) \propto N^{-\tau}$ (see Eq.(10.6)). Therefore the MSD can be evaluated by averaging Eq.(10.23) over the probability of finding the walker in a cluster of the corresponding size:

$$\overline{\langle \mathbf{r}_n^2 \rangle} \sim \sum_{N=1}^{n^{d_s/2}} N^{1/d_f+1-\tau} + \sum_{N=n^{d_s/2}}^{\infty} n^{1/d_w} N^{1-\tau}.$$

Here the first summand corresponds to the clusters on which the MSD has already stagnated, and the second one to those where it still grows. Approximating the sums by integrals we get

$$\overline{\langle \mathbf{r}_n^2 \rangle} \sim n^{1/d_w+(2-\tau)d_s/2}. \tag{10.24}$$

The same approach can be used to calculate the mean number of distinct sites visited. Since for percolation $d_s < 2$ in any spatial dimension d, we have

$$\langle S_n \rangle \sim \begin{cases} n^{d_s/2} \text{ for } n < N^{2/d_s} \\ N \quad \text{for } n > N^{2/d_s} \end{cases} \tag{10.25}$$

(the last expression corresponds to the case where all sites of the cluster are visited). The same averaging procedure as that leading to Eq.(10.24) would lead to

$$\overline{\langle S_n \rangle} \sim \sum_{N=1}^{n^{d_s/2}} N^{1+1-\tau} + \sum_{N=n^{d_s/2}}^{\infty} n^{d_s/2} N^{1-\tau} \sim n^{(3-\tau)d_s/2}, \tag{10.26}$$

which defines the "effective spectral dimension" $\tilde{d}_s = d_s(3-\tau)$ of this heterogeneous system. We note that, although the mean number of distinct visited sites is correctly reproduced by Eq.(10.26), this may not be immediately used even in considering the target reaction in the percolation system. The problem is that in some smaller clusters the reaction partners (B particles) may be simply missing, so that the A particles will survive indefinitely. The same is true for trapping. This means that the "effective spectral dimension" $\tilde{d}_s$ is not as universal as d_s and that for many quantities in heterogeneous systems the cluster averaging has to be performed at the last step of calculations. This can be seen when trying to calculate the effective spectral dimension $\tilde{\tilde{d}}_s$ not via $\langle S_n \rangle$ but according to its definition, Eq.(10.4).

Exercise 10.3 The spectral dimension of a structure is defined by the probability of being at the origin, that behaving as

$$P_n(0) \sim \begin{cases} n^{-d_s/2} & \text{for } n < N^{2/d_s} \\ N^{-1} & \text{for } n > N^{2/d_s} \end{cases}$$

on a finite cluster. Obtain the effective spectral dimension of the system via $\overline{P_n(0)} \propto n^{-\tilde{\tilde{d}}/2}$.

References

[1] B.D. Hughes. *Random Walks and Random Environments*, Vol. 2: *Random Environments*, Oxford: Clarendon Press, 1996
[2] D. Stauffer and A. Aharony. *Introduction to Percolation Theory*, London: Taylor and Francis, 1992
[3] J.M. Ziman. *Models of Disorder: The theoretical Physics of Homogeneously Disordered Systems*, Cambridge: Cambridge University Press, 1979
[4] R. Kopelman. *Science* **241**, 1620 (1988)
[5] Y. Meroz, I.M. Sokolov, and J. Klafter. *Phys. Rev. E* **81**, 010101 (2010)
[6] S. Alexander and R. Orbach. *Journal de Physique Lett.* **43**, L625 (1982)
[7] P. Grassberger. *Physica A* **262**, 251 (1999)
[8] B. O'Shaughnessy and I. Procaccia. *Phys. Rev. Lett.* **54**, 455 (1985)
[9] D.J. Frank and C.J. Lobb. *Phys. Rev. B* **37**, 302 (1988)

Further reading

B.B. Mandelbrot. *The Fractal Geometry of Nature*, New York: W.H. Freeman, 1982
T. Nakayama and K. Yakubo. *Fractal Concepts in Condensed Matter Physics*, Berlin: Springer, 2003
S. Havlin and D. ben Avraham. *Adv. in Physics* **51**, 187 (2002)
D. ben Avraham and S. Havlin, Diffusion and Reactions in Fractals and Disordered Systems, Cambridge: Cambridge Univercity Press, 2005

Index